RIGHTSTART™ MATHEMATICS

by Joan A. Cotter, Ph.D.

A HANDS-ON GEOMETRIC APPROACH LESSONS

Three-D images are made with Pedagoguery Software, Inc's Poly (http://www.peda.com/poly)

Printed in the United States of America

www.RightStartMath.com

For questions or for more information:
info@RightStartMath.com

To place an order or for additional supplies:
www.RightStartMath.com
order@RightStartMath.com

Activities for Learning, Inc.
PO Box 468; 321 Hill Street
Hazelton ND 58544-0468

888-272-3291 or 701-782-2002

701-782-2007 fax

ISBN 978-1-931980-38-8

April 2014

Table of Contents

Table of Contents

Table of Contents

Table of Contents

Vocabulary First Introduced

Lesson 1	line segments, parallel lines, intersection
Lesson 2	horizontal, vertical, diagonal, hexagon
Lesson 3	polygon, vertex, vertexes, vertices
Lesson 4	quadrilateral, equilateral triangle
Lesson 5	congruent
Lesson 6	bisect, tick mark, tetrahedron
Lesson 10	perimeter
Lesson 13	parallelogram
Lesson 14	rectangle, right angle, perpendicular
Lesson 15	rhombus
Lesson 16	90 degrees, square
Lesson 17	trapezoid, Venn diagram
Lesson 18	fraction
Lesson 19	numerator, denominator
Lesson 21	crosshatch
Lesson 23	ratio
Lesson 24	area, square centimeter
Lesson 25	area, square inch
Lesson 26	formula
Lesson 28	exponent
Lesson 30	factor
Lesson 32	millimeter, square millimeter
Lesson 34	little square, altitude
Lesson 36	isosceles
Lesson 38	distributive property, straightedge
Lesson 42	goniometer
Lesson 43	supplementary, vertical, complementary
Lesson 45	acute, obtuse, scalene
Lesson 46	external, internal, adjacent angle
Lesson 47	corresponding, alternate, interior, exterior angles
Lesson 48	SSS
Lesson 49	similar, SAS, ASA
Lesson 50	vertex angle, base angles, base
Lesson 51	median of a triangle
Lesson 52	centroid
Lesson 54	inscribed
Lesson 55	convex, concave
Lesson 56	hypotenuse, leg
Lesson 57	oblique
Lesson 58	Pythagorean theorem
Lesson 59	square root, integer, perfect square
Lesson 60	Pythagorean triple
Lesson 62	point, line, and plane, circumference, diameter, radius, arc, sector
Lesson 64	inscribed polygon, regular polygon
Lesson 65	tangent, tangent segment
Lesson 66	circumscribed polygon
Lesson 67	pi, π

Vocabulary First Introduced

Lesson 68   clockwise, counterclockwise
Lesson 69   concentric, semicircle
Lesson 70   internally tangent circles, externally tangent circles, trefoil, quatrefoil
Lesson 71   angle bisector, incenter
Lesson 72   chord, circumcenter*
Lesson 73   foot, feet
Lesson 74   central angle
Lesson 75   inscribed angle, intercepted arc
Lesson 76   kilometer
Lesson 80   per, unit cost
Lesson 81   tangram
Lesson 83   reflection, image, line of reflection, flip horizontal, flip vertical
Lesson 86   transformation
Lesson 87   translation, image, absolute, relative
Lesson 88   transformation
Lesson 93   angle of incidence, angle of reflection
Lesson 94   line of symmetry, maximum, minimum, ∞
Lesson 95   order of rotation symmetry, point symmetry
Lesson 97   frieze, cell, tile
Lesson 98   tessellation
Lesson 99   pure tessellation
Lesson 100   nonagon, decagon , dodecagon
Lesson 101   semiregular tessellation
Lesson 102   demiregular tessellation, semi-pure tessellation
Lesson 103   unit, pattern
Lesson 105   tartan, plaid, warp, weft, woof
Lesson 108   Escher
Lesson 109   Mondrian
Lesson 110   fractals and the terms iteration and self-similar, exponent
Lesson 111   Sierpinski Triangle
Lesson 112   Koch Snowflake
Lesson 114   similar, similar triangles
Lesson 115   proportion
Lesson 116   cross-multiplying
Lesson 118   golden rectangle, golden ratio, phi, φ
Lesson 119   golden spiral, golden triangle
Lesson 120   sequence, Fibonacci sequence
Lesson 121   Fibonacci spiral
Lesson 123   generalize
Lesson 129   Euler path
Lesson 131   trigonometry, opposite, adjacent, sine, cosine, tangent
Lesson 133   scientific calculator
Lesson 135   angle of elevation, stride, clinometer
Lesson 136   angle of depression
Lesson 137   sine wave

Vocabulary First Introduced

Lesson 138   solid, polyhedron, polyhedra, face, edge, vertex, net, dimension
Lesson 140   volume, cubic centimeter, surface area
Lesson 141   decimeter, dm
Lesson 142   prism
Lesson 143   short diagonal, long diagonal
Lesson 144   cylinder
Lesson 145   cone
Lesson 146   apex, regular pyramid, right pyramid
Lesson 149   Platonic solids
Lesson 151   dual polyhedra
Lesson 152   sphere, great circle, small circle
Lesson 153   planes of symmetry
Lesson 154   axes of symmetry
Lesson 155   reciprocal
Lesson 157   stella octangula, concave polyhedron
Lesson 158   truncate, semiregular polyhedra, Archimedean solids
Lesson 159
Lesson 160
Lesson 161
Lesson 162
Lesson 163   quasiregular polyhedron
Lesson 164
Lesson 165

Photo Credits

Lesson 3	Irene Genelin, Andy Cotter
Lesson 4	Kathleen Lawler
Lesson 8	Public Domain
	Public Domain
	Irene Genelin
Lesson 15	Irene Genelin
Lesson 30	Irene Genelin
Lesson 31	Irene Genelin
Lesson 36	Kathleen Lawler
Lesson 40	Irene Genelin
Lesson 44	Alvin Cotter
Lesson 61	Irene Genelin
Lesson 65	Irene Genelin
Lesson 69	Irene Genelin, Andy Cotter
Lesson 70	Irene Genelin
Lesson 71	Irene Genelin
Lesson 74	Irene Genelin
Lesson 75	Irene Genelin
Lesson 76	Irene Genelin
Lesson 78	Irene Genelin, Andy Cotter
Lesson 83	Irene Genelin
Lesson 94	Irene Genelin
Lesson 95	Irene Genelin
Lesson 121	Irene Genelin
Lesson 136	Irene Genelin
Lesson 141	Irene Genelin
Lesson 144	Irene Genelin
Lesson 145	Irene Genelin
Lesson 146	Irene Genelin, Andy Cotter
Lesson 160	Irene Genelin, Andy Cotter

RightStart™ Mathematics: A Hands-On Geometric Approach

RightStart™ Mathematics: A Hands-On Geometric Approach is an innovative approach for teaching many middle school mathematics topics, including perimeter, area, volume, metric system, decimals, rounding numbers, ratio, and proportion. The student is also introduced to traditional geometric concepts: parallel lines, angles, midpoints, triangle congruence, Pythagorean theorem, as well as some modern topics: golden ratio, Fibonacci numbers, tessellations, Pick's theorem, and fractals. In this program the student does not write out proofs, although an organized and logical approach is expected.

Understanding mathematics is of prime importance. Since the vast majority of middle school students are visual learners, approaching mathematics through geometry gives the student an excellent way to understand and remember concepts. The hands-on activities often create deeper learning. For example, to find the area of a triangle, the student must first construct the altitude and then measure it. If possible, students work with a partner and discuss their observations and results.

Much of the work is done with a drawing board, T-square, 30-60 triangle, 45 triangle, a template for circles, and goniometer (device for measuring angles). Constructions with these tools are simpler than the standard Euclid constructions. It is interesting to note that CAD (computer aided design) software is based on the drawing board and tools.

This program incorporates other branches of mathematics, including arithmetic, algebra, and trigonometry. Some lessons have an art flavor, for example, constructing Gothic arches. Other lessons have a scientific background, sine waves, and angles of incidence and reflection; or a technological background, creating a design for car wheels. Still other lessons are purely mathematical, Napoleon's theorem and Archimedes stomachion. The history of mathematics is woven throughout the lessons. Several recent discoveries are discussed to give the student the perspective that mathematics is a growing discipline.

Good study habits are encouraged through asking the student to read the lesson before, during, and following the worksheets. Learning to read a math textbook is a necessary skill for success in advanced math classes. Learning to follow directions is a necessary skill for studying and everyday life. Occasionally, an activity or lesson refers to previous work making it necessary for the student to keep all work organized. The student is asked to maintain a list of new terms.

This text was written with several goals for the student: a) to use mathematics previously learned, b) to learn to read math texts, c) to lay a good foundation for more advanced mathematics, d) to discover mathematics everywhere, and e) to enjoy mathematics.

About the author
Joan A. Cotter, Ph.D., author of *RightStart™ Mathematics: A Hands-On Geometric Approach* and *RightStart™ Mathematics* elementary program has a degree in electrical engineering, a Montessori diploma, a masters degree in curriculum and instruction, and a doctorate in mathematics education. She taught preschool, children with special needs, and mathematics to grades 6-8.

Hints on Tutoring
RightStart™ Mathematics: A Hands-On Geometric Approach

Before starting a lesson, the student should look over the Materials list and gather all the supplies, including a mechanical pencil or a sharp #2 pencil and a good eraser. Then the student reads over the goals, keeping in mind that italicized words will be explained in the lesson. (These new words are to be recorded in the student's math dictionary.) Next the student begins reading the Activities, carefully studying any accompanying figures. It is a good habit to summarize the activity after reading it. If a paragraph is unclear, the student should reread the paragraph, keeping in mind that sometimes more is explained in the following paragraph. No one learns mathematics by reading the text only once.

Each activity needs to be understood before going to the next activity. Make sure the student understands the importance of completing the problems on the worksheet when called for in the lesson. Sometimes it will be necessary to refer to the lesson while completing the worksheet. All work needs to be kept neatly in a three-ring binder for future reference.

Be careful how you react to the "I don't get it" plea. Tell the student you need a question to answer. You do not want to get in the habit of reading the text for your student and then regurgitating to her like a mother robin feeding her young. The text is written for students to read for themselves. Learning how to ask questions is an important skill to acquire toward becoming an independent learner. If questions remain after diligent study, the student can contact the author at JoanCotter@RightStartMath.com.

If a child has a serious reading problem, read the text aloud while he follows along and then ask him to read it aloud. Be sure each word is understood. For less severe reading problems, you might model aloud the process of reading an activity, commenting on the figure, and summarizing the paragraph. Some of the time, students need encouragement to overcome frustration, which is inherent in the learning process. Occasionally, a student may have a knowledge gap needed for a particular lesson, requiring other resources to resolve. Incidentally, research shows one of the major causes of difficulties in learning new concepts for this age group is insufficient sleep.

After the student has completed the worksheet, ask her to compare her work with the solution. If the student has a partner, they can compare and discuss their work before referring to the solutions. Ask her to explain what she learned and any discrepancies. Keep in mind that some activities have more than one solution. You might also ask her to grade her work on some agreed upon scale. It also is a good idea for the student to reread the goals of the lesson to see if they have been met. Encourage discussion on practical applications of the topic.

Lesson 1   **Getting Started**

GOALS   1. To learn to tape the paper to the drawing board
2. To learn to draw lines with a T-square and with a triangle
3. To identify *line segments, parallel lines, and intersections*

MATERIALS   A blank sheet of paper, 3M Removable Tape, Worksheet 1
Drawing board, T-square, 30-60 triangle
Sharp pencil (preferably mechanical 0.5) and eraser
Math Dictionary (page 1) found before Worksheet 1

PREPARATION   ***Taping the paper.*** Position the drawing board as shown below.
Tape the paper to the board only at the top two corners.

> ***Do not tape the lower corners. If you do, the tape will interfere with the T-square.***

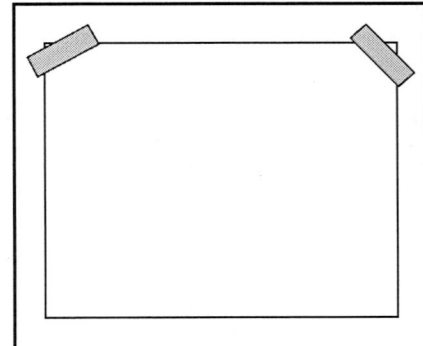

Align the paper and tape it to the drawing board. Tape only the top corners.

The tape will not be shown again.

ACTIVITIES   ***Horizontal lines.*** Use the T-square to draw horizontal lines. A right-handed person places the T-square along the left side of the board. A left-handed person places it along the right side.

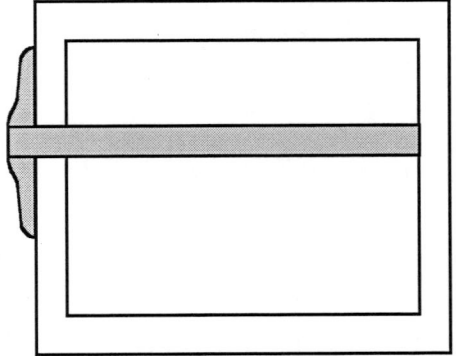

Position of the T-square if you are right-handed.

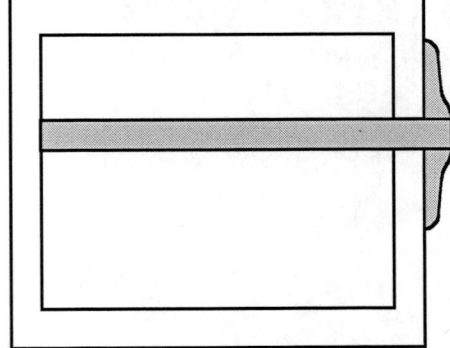

Position of the T-square if you are left-handed.

> ***Keep your T-square along the side of your board. Draw lines only along the upper edge.***

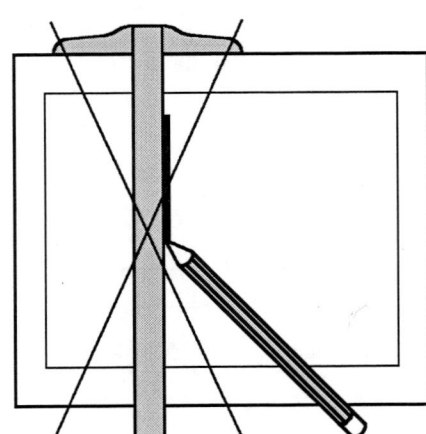

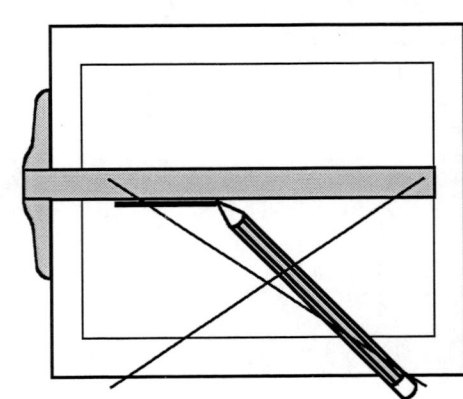

Practice drawing some horizontal lines. Before drawing each line, check to be sure your T-square is against the side of your drawing board. Leave margins on all four sides of your paper.

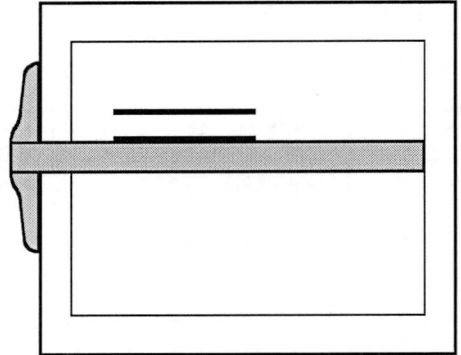

Drawing horizontal lines.

Drawing vertical lines. To draw vertical lines, place the 30-60 triangle on the T-square. Be sure the triangle is against the T-square and the T-square is against the board before starting to draw the line. Start the line about 1 cm (a half inch) from your T-square. Slide the triangle to draw the next line. Draw only on the outside of the triangle.

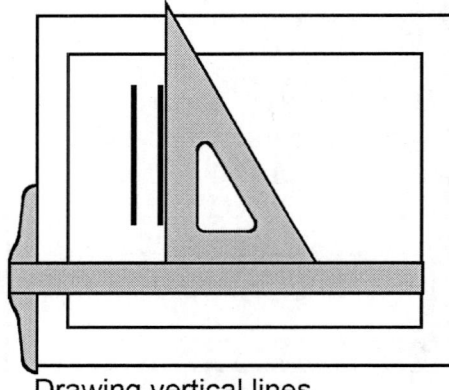

Drawing vertical lines.

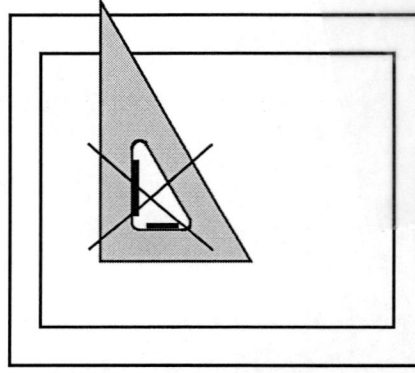

Do not draw inside the triangle.

Drawing some other lines. Draw lines using the other side of the triangle. Also turn your triangle as shown below and draw more lines as shown in the figures below and on the next page.

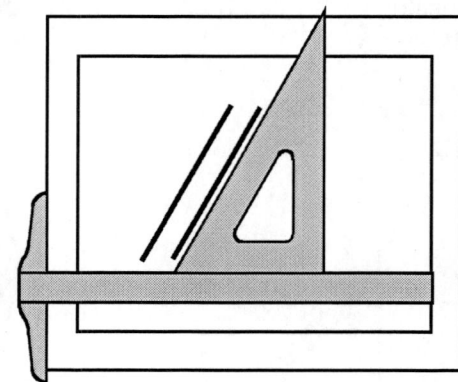

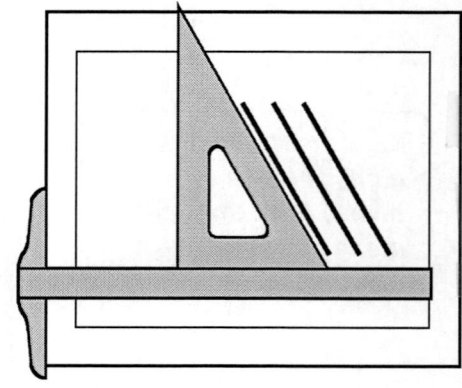

Drawing lines with other edges of the 30-60 triangle.

Many lines connect the star to the edges of the door.

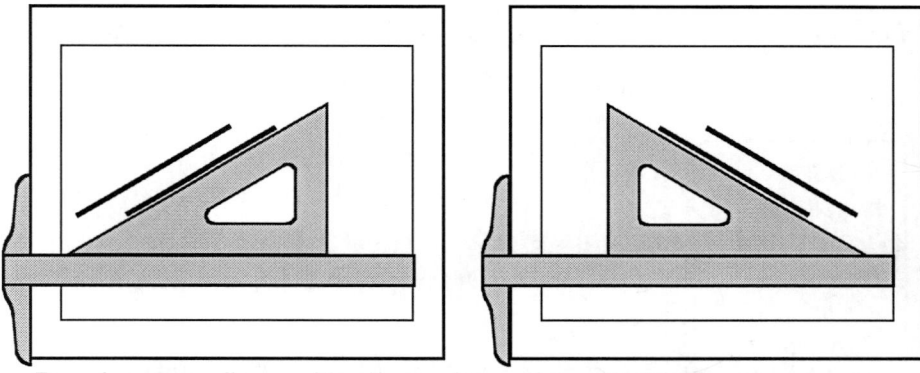

Drawing more lines with other edges of the 30-60 triangle.

Lines and line segments. The word *line* has many meanings. One dictionary gives 35 definitions, including a line of scrimmage. It also can mean a wrinkle or the division between two states or countries.

To mathematicians a line has no ends. It travels forever in both directions. When they talk about part of a line with definite ends, mathematicians use the term *line segment*.

Parallel lines. You must have noticed that the line segments were parallel when you were drawing with only the T-square and when sliding the triangle on the T-square. The usual definition of parallel lines is that they never meet. (Apparently, they're not very friendly.) At every point on the lines the distance between them is the same.

Intersecting lines. Lines that do meet are called *intersecting lines*. Where two streets meet is an intersection. The word intersect comes from *inter* meaning between and *sect* meaning to cut. Can parallel lines be intersecting lines? The answer is at the bottom of the page.

Worksheet. This worksheet is similar to the exercises suggested in this lesson. To align your paper on the drawing board, place it on the board. Next place the T-square next to a horizontal line. Adjust the paper until the line on the paper is parallel to the T-square. Then affix the tape. See the figures below.

> *Keep your worksheets in a binder. From time to time you will need to refer to them.*

Brick design in a wall.

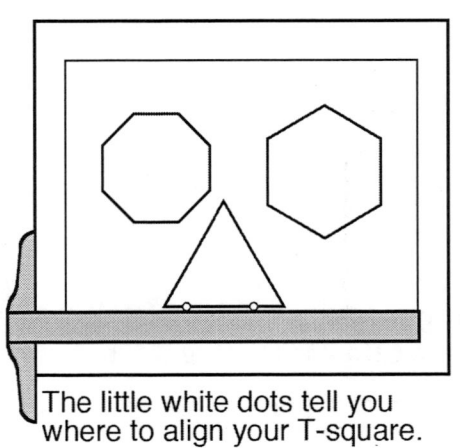

The little white dots tell you where to align your T-square.

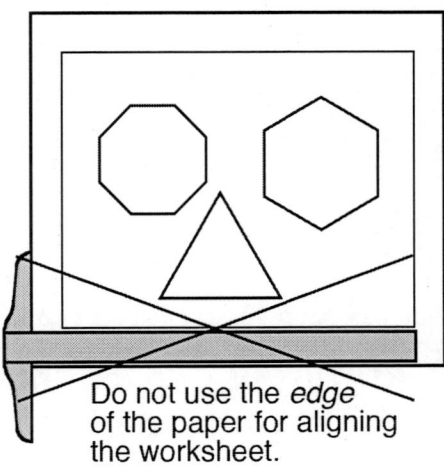

Do not use the *edge* of the paper for aligning the worksheet.

Math dictionary. Write definitions for the words listed at the beginning of the lesson under Goal 3. Use your own words to write the definitions. [Answer: no]

4

Drawing Diagonals

<table>
<tr><td>**GOALS**</td><td>1. To review the terms *horizontal* and *vertical*
2. To learn the mathematical meaning of *diagonal*
3. To review the term *hexagon*
4. To find the correct edge of the 30-60 triangle to draw diagonals</td></tr>
</table>

A sharp pencil, an eraser, and tape are essentials. They will not be listed in future lessons.

MATERIALS Worksheet 2, Math Dictionary
Drawing board, T-square, 30-60 triangle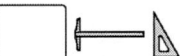

ACTIVITIES ***Horizontal and vertical.*** *Horizontal* refers to the horizon, the intersection between the earth and sky. You can see it if there aren't too many buildings and trees in the way. *Vertical* refers to straight up and down, like a flagpole.

A *horizontal* line on paper is a line drawn straight across the paper. It usually is parallel to the top and bottom of the paper. A *vertical* line on paper goes from top to bottom, parallel to the sides of the paper.

Diagonals. In common everyday English, the word *diagonal* usually means at a slant. It often means a road that runs neither north and south nor east and west.

In mathematics, a *diagonal* is a line connecting points in a closed figure. For example, the line segments $\overline{AC}$ and $\overline{DB}$ drawn in the square below on the left are diagonals. If we turn the square, as in the next figure, the lines segments are still diagonals. Now diagonal $\overline{DB}$ is horizontal and diagonal $\overline{AC}$ is vertical.

Diagonal lines on a building.

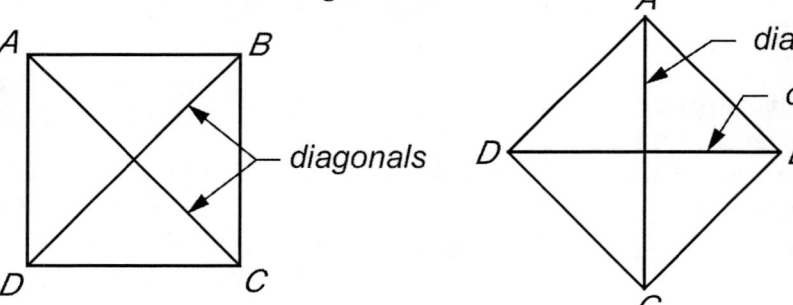

In the word diagonal, *dia* means "across" and *gon* means "angle." So, a diagonal is a line across angles, that is, a line connecting two vertices.

Worksheet. The worksheet asks you to draw two hexagons and all their diagonals. A *hexagon* is a closed six-sided figure. One way to remember the word is that *hexagon* and *six* both have *x*'s.

Draw the sides of the hexagon and the diagonals using your tools. The horizontal and vertical lines need only a T-square. The left figure below is a hexagon; the right figure shows the diagonals.

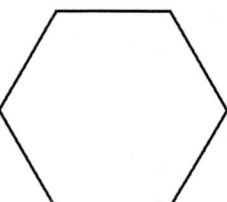
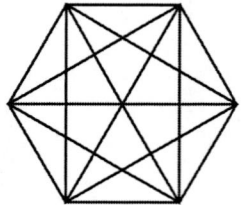

Lesson 3

Drawing Stars

GOALS   1. To learn the term *polygon*
2. To learn the term *vertex* and its plurals, *vertices* and *vertexes*
3. To draw stars by following instructions shown in pictures

MATERIALS   Worksheets 3-1 and 3-2
Math Dictionary
Drawing board, T-square, 30-60 triangle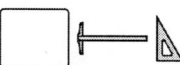

ACTIVITIES   ***Polygons.*** In the row below are examples of figures that are polygons and figures that are not polygons. Before reading further, think of a good definition for polygon.

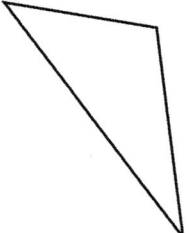 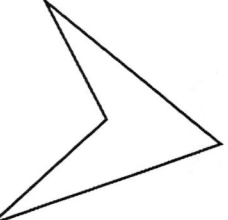 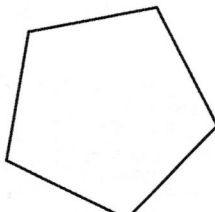

These are polygons.

 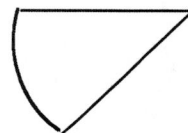 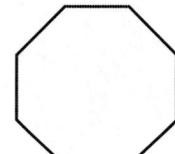

These are NOT polygons.

Did your definition include a closed figure with straight line segments?

Vertex and its plurals. In a polygon the point where the lines meet is call a vertex. You have two choices for the plural of *vertex*, either *vertices* (VER-ti-sees) or *vertexes*. For some reason, even though the word *vertexes* follows the normal English rule for plurals, math books (and tests) prefer *vertices*.

For example, there are three vertices in a triangle and four vertices in a square.

Worksheets. On the worksheets, you will be drawing stars. The boldfaced lines in the little figures tell you what to draw. Be sure to use your T-square and 30-60 triangle to draw all the lines. (You need only your T-square for the horizontal lines.) The completed stars will look like the figures below.

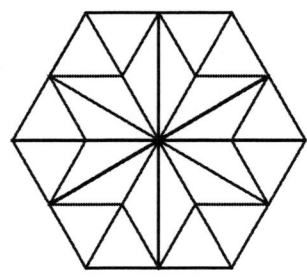 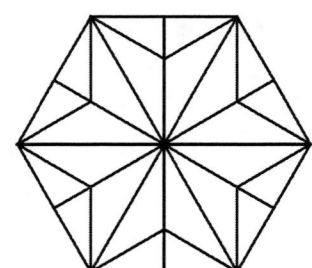

Star designs in Morocco, where they are very common.

Lesson 4

Equilateral Triangles into Halves

GOALS
1. To review the terms *quadrilateral* and *equilateral triangle*
2. To draw equilateral triangles
3. To divide equilateral triangles into halves

MATERIALS
Worksheet 4, Math Dictionary
Drawing board, T-square, 30-60 triangle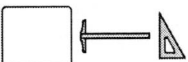

ACTIVITIES
Quadrilaterals and equilateral triangles. The following figures are all quadrilaterals.

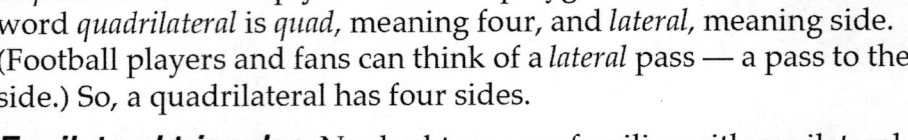

A *quadrilateral* is simply a four-sided polygon. The derivation of the word *quadrilateral* is *quad*, meaning four, and *lateral*, meaning side. (Football players and fans can think of a *lateral* pass — a pass to the side.) So, a quadrilateral has four sides.

Equilateral triangles. No doubt you are familiar with equilateral triangles. To refresh your memory, several are shown below.

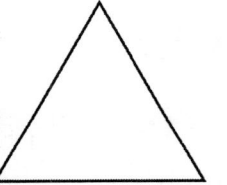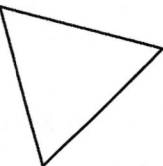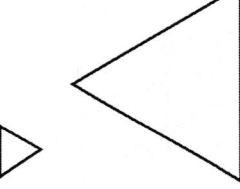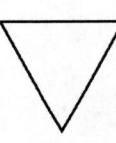

A triangle is a three-sided polygon. The meaning of *equilateral* is *equi*, meaning *equal*, and *lateral*, meaning, of course, *side*. So, an equilateral triangle has equal sides.

Drawing an equilateral triangle. To draw a side of the equilateral triangle, slide your T-square down 1 cm ($\frac{1}{2}$ inch) below the base line. Then use the 60° angle of your triangle. See the figures below.

The building has a pattern of quadrilaterals.

This stained glass window contains many quadrilaterals

> *Your triangle needs to be below the the line you're drawing.*

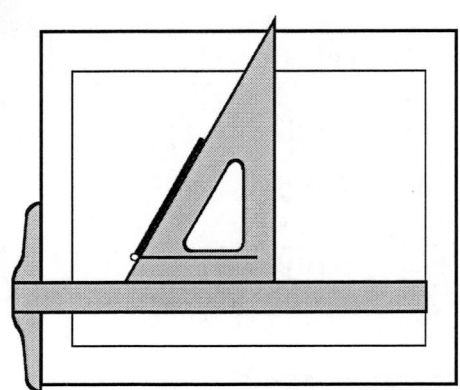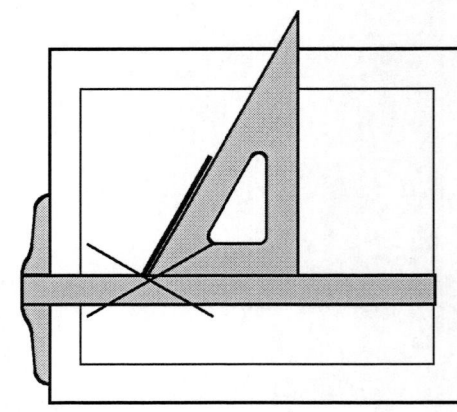
Drawing a side of the equilateral triangle.

Where two pyramids almost touch in the Louvre (LOOV-ruh) Museum, Paris, France.

Building in England adorned with equilateral triangles.

The body of this moth is an equilateral triangle.

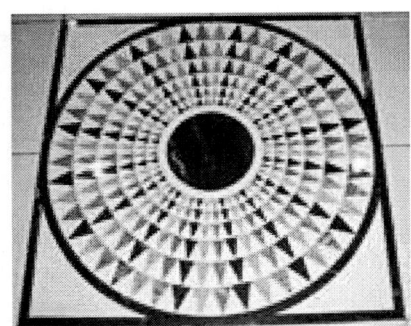

Many equilateral tirangles in this design on a floor.

A problem arises because you don't know how long to draw the line. Just draw it a little longer than you think you need — that's why you have an eraser.

The next step is to flip your triangle over. Then draw the second side of the triangle starting at the other end of the base line. See the figure below. This time you will know how long to draw the line.

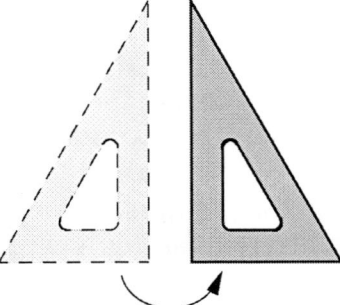

Flipping a triangle.

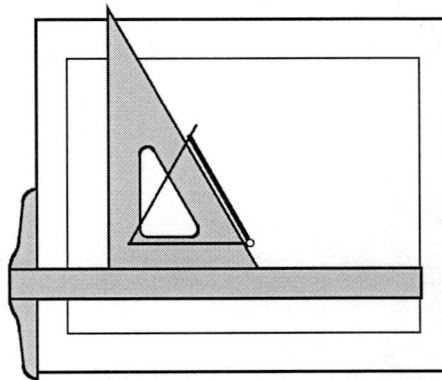

Drawing the third side of the equilateral triangle.

If you discover you didn't draw the first line long enough, use your T-square and triangle to draw it longer. Don't cheat and draw it with just your pencil.

Half of an equilateral triangle. Imagine cutting out an equilateral triangle and folding it in half. A triangle can be folded in half three different ways. As an example, see the figures below. Notice that the line dividing the triangle in half is drawn between a vertex and the *midpoint*, or middle, of the opposite side.

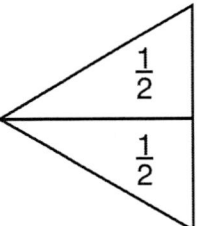
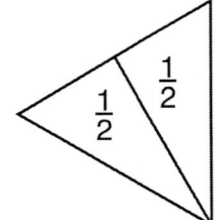
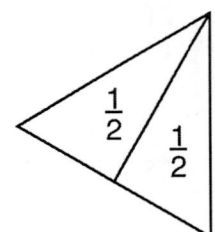

Worksheet. For this worksheet you are to draw six equilateral triangles. One line is given for each triangle. The miniature triangles show what they should look like.

After you have drawn the equilateral triangles, divide them into two equal parts, again using your tools. Lastly, write $\frac{1}{2}$ on each half.

Lesson 5

Equilateral Triangles into Sixths & Thirds

GOALS
1. To review the term *congruent*
2. To learn the term *centroid* for an equilateral triangle
3. To divide equilateral triangles into sixths and thirds
4. To find the balance point of an equilateral triangle

MATERIALS
Worksheet 5, Math Dictionary
Drawing board, T-square, 30-60 triangle
Scissors, pencil with new eraser

ACTIVITIES

Congruent triangles. In the previous lesson, you divided an equilateral triangle into two identical parts. If you cut them out, you could lay one part exactly on top of the other. Try to do this mentally with the of figures below. Notice that you must flip one part over. When two figures line up exactly, they are *congruent*.

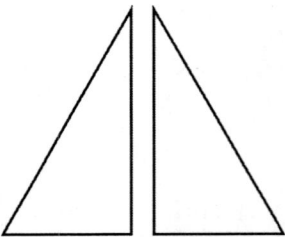
Flip one triangle before stacking to see them as congruent.

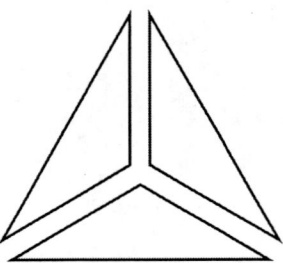
Stack the three triangles in your mind to see them as congruent.

> When you divide 1 into 6 equal parts, each part is 1 ÷ 6, usually written as $\frac{1}{6}$. The bar in the middle means divided by.

Centroid of an equilateral triangle. The figure below shows an equilateral triangle divided into halves three ways. The point where the lines intersect is the *centroid*.

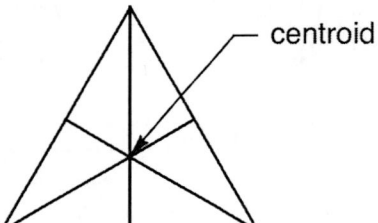
centroid

Dividing a triangle into 6ths and 3rds. On the worksheet, you're asked to divide equilateral triangles into sixths and thirds. See the figures below. Notice that thirds are like sixths with parts of the lines missing.

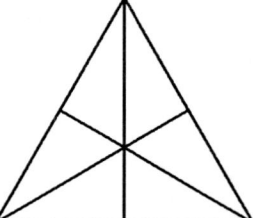 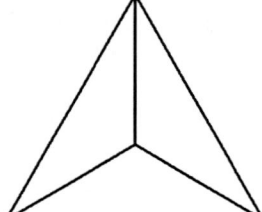 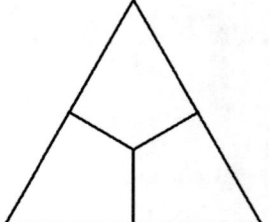

Balance point of a triangle. The last project on the worksheet asks you to find the point where an equilateral triangle will balance. First guess and then check using the pencil with a new eraser.

Lesson 6

Equilateral Triangles into Fourths & Eighths

GOALS
1. To learn the term *bisect*
2. To learn to draw *tick marks*
3. To construct a *tetrahedron*
4. To divide equilateral triangles into eighths

MATERIALS
Worksheet 6, Math Dictionary
Drawing board, T-square, 30-60 triangle, scissors

ACTIVITIES

Dividing a triangle into fourths. An equilateral triangle divided into fourths is shown at the right.

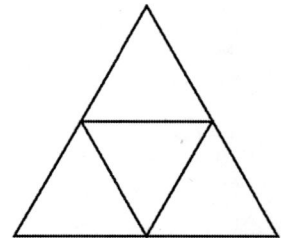

Construct it by first *bisecting* one of the sides, that is, finding its midpoint. See the left figure below. Do this by preparing to divide the triangle in half. Then instead of drawing the entire line segment, draw only a short line. This line is called a *tick mark*.

> **Be sure to extend the tick mark on both sides of the line. Then you will have a good point for the next line.**

> **Do not erase your tick marks. They provide a record of your work.**

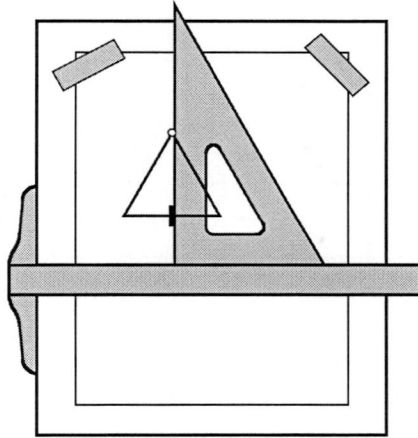

Bisecting the base line with a tick mark.

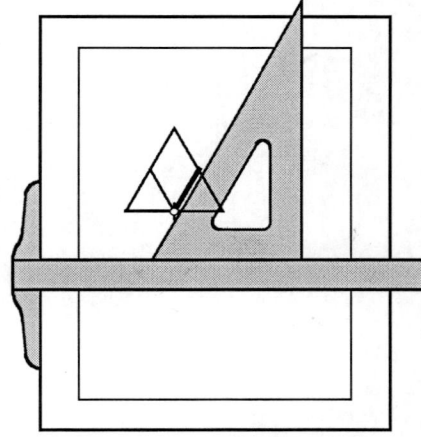

Drawing a line starting at the tick mark.

The figure above on the right shows drawing the next two lines. Only the horizontal line is left to draw.

A different paper orientation. Notice that the worksheet for this lesson is in the "portrait" position. (The other direction is called "landscape." The short sides are at the top and bottom. Arrange your drawing board as shown above. Place the tape at the top.

Making a tetrahedron. On the worksheet draw an equilateral triangle and divide it into fourths. Cut it out and fold on the lines so the vertices of the original triangle touch. See the figure shown on the right.

This object is called a *tetrahedron* (TE-truh-HEE-drun), which is a special pyramid. The word *tetrahedron* is Greek for four, *tetra*, and faces, *hedron*. How many faces, or surfaces, does the tetrahedron have?

Lesson 7

Equilateral Triangles into Ninths

GOALS
1. To divide an equilateral triangle into ninths
2. To observe a pattern and to continue it

MATERIALS
Worksheet 7, Math Dictionary
Drawing board, T-square, 30-60 triangle

ACTIVITIES
Dividing a triangle into ninths.
An equilateral triangle divided into ninths is shown at the right.

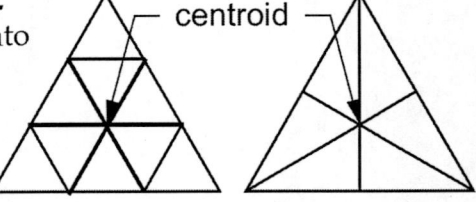

Observe that three lines are drawn through the centroid and parallel to the sides. So, you must first find the centroid. Do this with tick marks as shown below. Then draw the remaining lines.

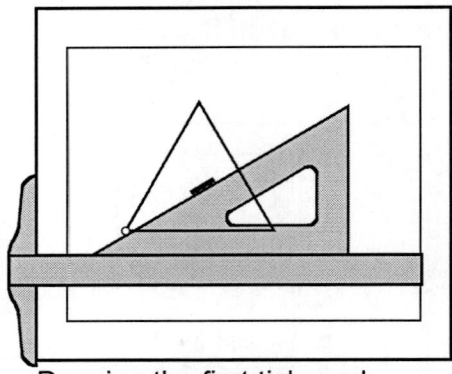
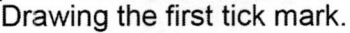

Drawing the first tick mark.

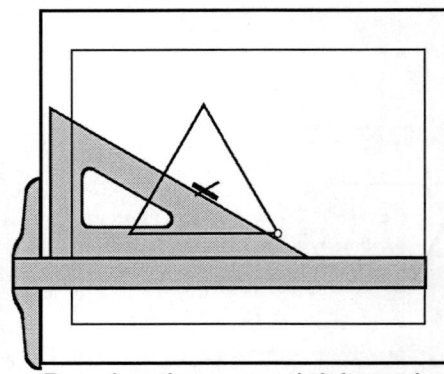

Drawing the second tick mark.

Describing the pattern. This figure has an interesting pattern. See the left figure below, which has an extra row. Observe the number of small triangles in each row. The worksheet has a chart to help you see the results mathematically. It is shown below on the right.

The first column has the row number. The second column is twice the first column. The third column is the number of little triangles in each row.

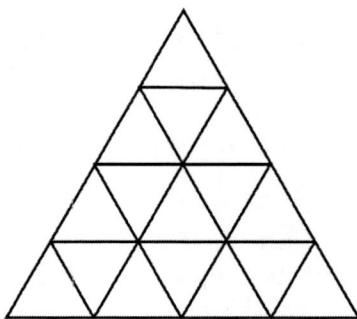

Row Num	Row Num × 2	Eq. Tri in Row
1	2	1
2	4	3
3		
4		
5		
6		
10		
n		

In the last row, n means any number. Twice any number is $2n$. Also, any number plus 1 is $n + 1$; any number minus 1 is $n - 1$.

Compare the second and third columns. Use this pattern to fill in the remaining columns.

Lesson 8

Hexagrams

GOALS 1. To compare how many times greater one figure is to another
2. To construct hexagrams in different ways

MATERIALS Worksheets 8-1, 8-2
Drawing board, T-square, 30-60 triangle
Scissors, glue, tape

ACTIVITIES ***Times greater.*** On the right are two equilateral triangles, one shaded. How many times greater is the larger triangle compared to the shaded triangle? One way to think about this is to find how many times the shaded triangle would fit, or fill up, the larger triangle. In this example, the answer is 4.

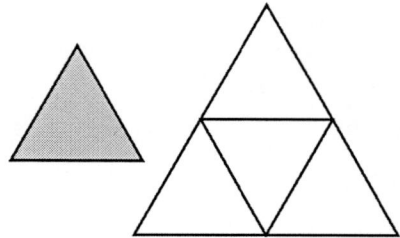

Now look at the three figures below.
How many times greater is the figure above compared to the shaded triangle below? Do not be fooled by extra lines. The answers are at the bottom of the page.

A window detail from Temple Square in Salt Lake City, Utah.

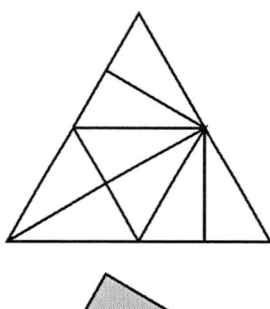 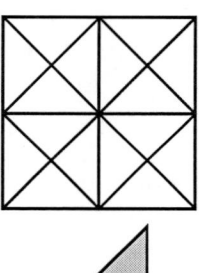 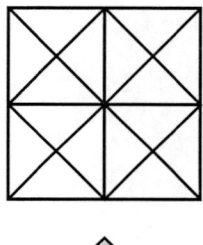

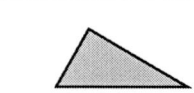

How many times greater is the figure above compared to the shaded triangle below?

Worksheets. On these two worksheets you are to construct hexagrams, a special six-sided star, by using three different methods. On the second worksheet, you will construct a variation of a hexagram, Solomon's seal.

Adrian, 11, discovered that you can see a star pattern when you hold your woven hexagram up to the light. [Answers: 8 times, 8 times, 16 times]

A synagogue, built in 1971, in Karlsruhe, Germany.

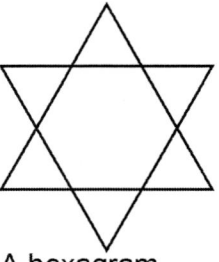

A hexagram.

Woven hexagram.

American cemetery in Normandy, France.

Lesson 9

Equilateral Triangle into Twelfths and More

GOALS
1. To discover how to divide an equilateral triangle into congruent pieces greater than 9
2. To divide an equilateral triangle into twelfths
3. To divide an equilateral triangle into a number greater than 12

MATERIALS
Worksheets 9-1, 9-2
Drawing board, T-square, 30-60 triangle
Colored pencil, optional

ACTIVITIES
Dividing a triangle into twelfths. How would you divide an equilateral triangle into twelfths – into twelve congruent parts? Think about it for a while before reading further. Would it work to divide the triangle into thirds and divide each third into fourths? One student even suggested dividing the triangle into tenths and then dividing each tenth in half. Let's hope he was joking!

If you have thought about it, you probably realize you first divide the triangle into fourths and then each fourth into thirds.

Dividing a triangle by higher numbers. How would you divide the triangle into sixteenths? What other numbers could you divide it into? Two kindergarten girls divided the equilateral triangle into 256 equal parts. After hearing about the girls, a teacher learning drawing board geometry divided his triangle into 432 equal parts. Some divisions are shown below.

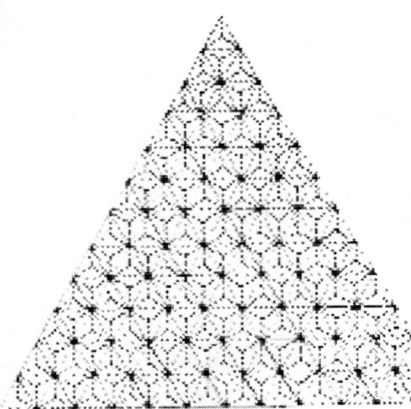

Triangle into 432nds by
Joseph Hermodson-Olsen, 14.

How could he have done it?
The answer is at the bottom of
the page.

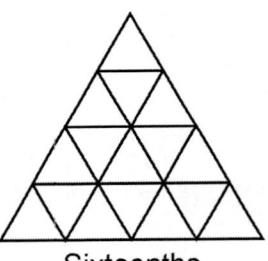

Sixteenths

Eighteenths

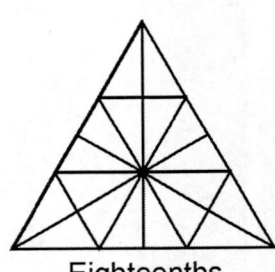

Eighteenths

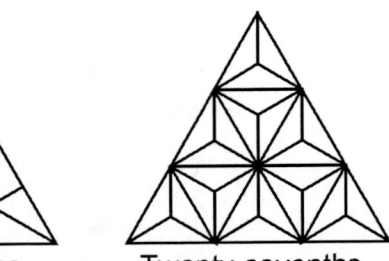

Twenty-fourths

Twenty-sevenths

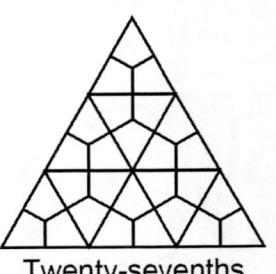

Twenty-sevenths

Worksheet 9-1. For this worksheet, you will divide the equilateral triangle into twelfths. Work carefully. For Problem 2, figure out how you would divide equilateral triangles into various congruent pieces.

Worksheet 9-2. After drawing the equilateral triangle, divide it into congruent triangles. Either copy one of the designs above, or better yet, design your own. You might like to color your design. [Answer: ninths, fourths, fourths, and thirds.]

Thirty-seconds

Lesson 10

Measuring Perimeter in Centimeters

GOALS 1. To construct a ruler in centimeters from the centimeter cubes
2. To measure *perimeter* in centimeters

MATERIALS Worksheet 10
Drawing board, T-square, 30-60 triangle, 10 centimeter cubes
4-in-1 ruler

ACTIVITIES **Measuring with tiles.** One way to measure the length of the left line shown below is to use the edge of centimeter cubes. Each edge is 1 cm. Lay them side to side as shown below on the right.

> The abbreviation for centimeter is <u>cm</u>. It has two lower-case letters with no period after it.

How long is this line segment?

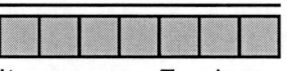
It measures 7 cubes.

Problem 2, constructing the centimeter ruler. You could do all your measuring using cubes, but it would quickly become tedious. So rulers were invented.

To construct the ruler on the worksheet, first line up the cubes. Then mark the centimeters with tick marks. Use your tools and extend each line to the bottom of the rectangle. See figures below.

Aligning the tiles.   Marking the centimeters.   Extending the marks.

Number your centimeters. Then divide each centimeter in half. Do it the same way as for bisecting a line in a triangle. See below.

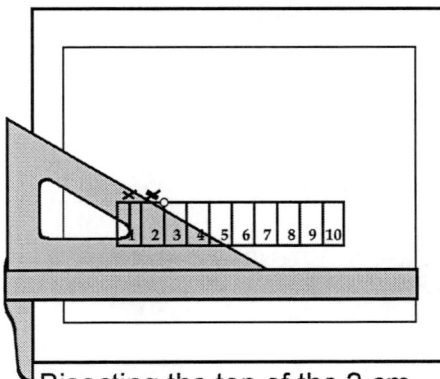
Bisecting the top of the 2 cm line to find the half mark.

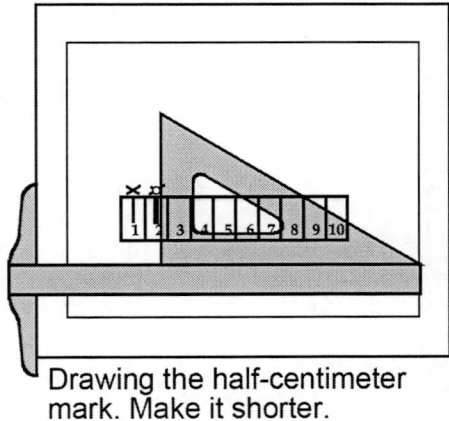
Drawing the half-centimeter mark. Make it shorter.

The centimeter ruler completed.

> See the word *"rim"* in pe**RIM**eter in order to remember perimeter measures the edges.

Worksheet. The word *perimeter* comes from two words; *peri-* means *all around* as in *periscope* and *meter* means *measure*. To find the perimeter of a figure, you measure all around it.

Use your 4-in-1 ruler to measure the quadrilaterals. One side of the ruler is metric and the other side is inches. Use the following format:

$$P = 2 + 3 + 2 + 3 \text{ (or } P = 2 \times 2 + 3 \times 2)$$
$$P = 10 \text{ cm}$$

Lesson 11

Drawing Parallelograms in Centimeters

A tapistry with parallelograms.

> *Dimensions are simply measurements, such as length and width.*

GOALS
1. To learn the definition for *parallelogram*
2. To draw parallelograms by measuring sides
3. To draw parallelograms with a given perimeter

MATERIALS
Worksheet 11
Drawing board, T-square, 30-60 triangle
4-in-1 ruler

ACTIVITIES
Parallelogram. The derivation of *parallelogram* is *parallel* and *gram*. *Gram*, as well as the word *graphic*, means line. A *parallelogram* is a quadrilateral with two sets of parallel lines. See examples below.

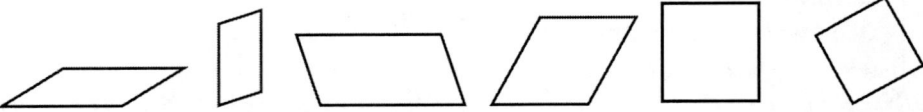

Drawing a parallelogram to given dimensions. To draw a parallelogram that has sides of 6 cm and 8 cm, use this procedure. Start by drawing a base line slightly longer than 8 cm. Then slide your T-square down and place your ruler under the line. Measure the 8 cm and draw a tick mark. See the left figure below.

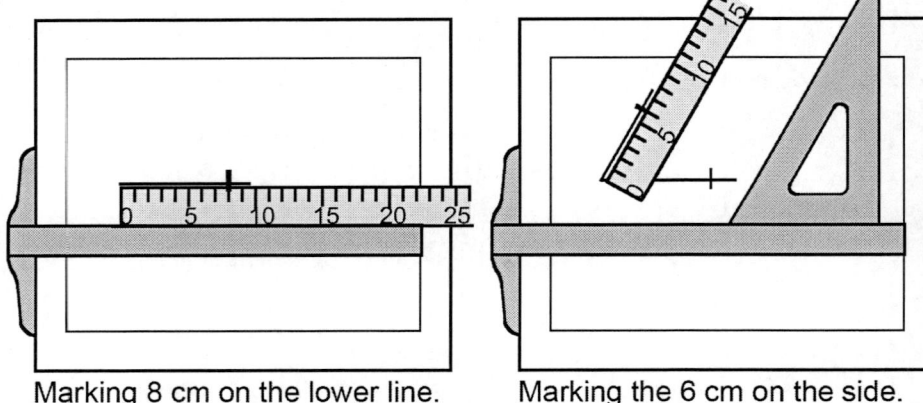

Marking 8 cm on the lower line.   Marking the 6 cm on the side.

Next use your triangle to draw a side and mark the 6 cm. See the right figure above. Complete the remaining sides as shown below.

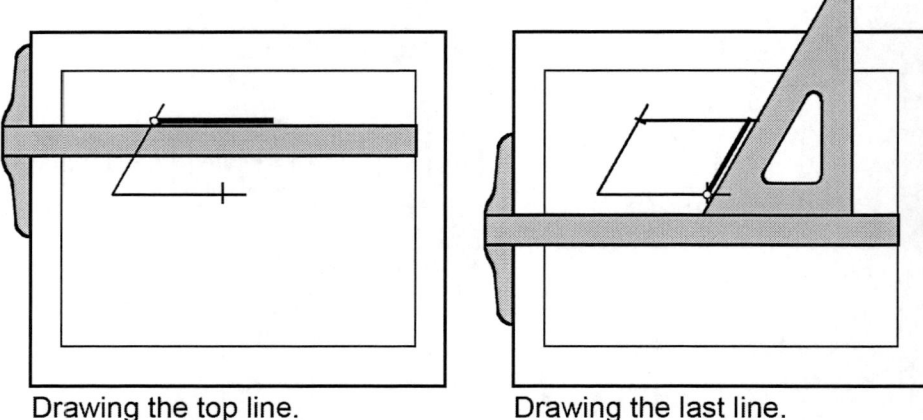

Drawing the top line.   Drawing the last line.

Worksheet. Draw three different parallelograms (you choose the shape) with perimeters equal to 18 cm.

Lesson 12, optional

Measuring Perimeter in Inches

> This lesson is very similar to Lesson 10. It may be omitted, especially if you don't use inches.

GOALS
1. To construct a ruler in inches from the tiles
2. To measure *perimeter* in inches

MATERIALS
Worksheet 12
Drawing board, T-square, 30-60 triangle, six 1" tiles
4-in-1 ruler

ACTIVITIES
Measuring with tiles. The edge of each tile should be 1 in. Measure how long the line below is in tiles. Do this by laying the tiles side to side as shown below.

This line measures 4 tiles long.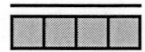

Problem 2, constructing the inch ruler. You could do all your measuring using tiles, but it would quickly become tiresome. So rulers were invented.

To construct the ruler on the worksheet, first line up the tiles. Then mark the inches with tick marks. Use your tools and extend each line to the bottom of the rectangle. See the figures below.

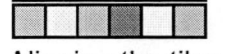
Aligning the tiles.

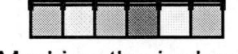
Marking the inches.

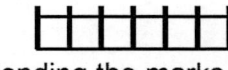
Extending the marks.

Next number your inches. Then divide each inch in half. Use the same method you used for bisecting a line in a triangle. See below.

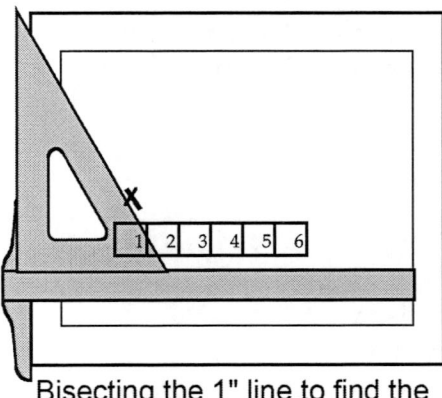

Bisecting the 1" line to find the half mark for the inch.

Drawing the half-inch mark. Make it shorter.

The inch ruler completed.

> See the word "rim" in peRIMeter in order to remember perimeter measures the edge.

Worksheet. The word *perimeter* comes from two words. *Peri-* means *all around* as in *periscope* and *meter* means *measure*. To find the perimeter of a figure, you need to measure the distance around it.

Use your 4-in-1 ruler to find the perimeter of the three rectangles on the worksheet. Use the following format. (If you prefer, use "in." instead of the symbol (") for inches.)

$$P = 2" + 3" + 2" + 3" \text{ or } (P = 2 \times 2 + 3 \times 2)$$
$$P = 10"$$

Lesson 13, optional

Drawing Parallelograms in Inches

> **This lesson is very similar to Lesson 11. It may be omitted, especially if you don't use inches.**

GOALS
1. To learn the definition for *parallelogram*
2. To draw parallelograms by measuring sides
3. To draw parallelograms with a given perimeter

MATERIALS
Worksheet 13
Drawing board, T-square, 30-60 triangle
4-in-1 ruler

ACTIVITIES
Parallelograms. The derivation of *parallelogram* is *parallel* and *gram*, which means *lines*, as does the word *graphic*. A *parallelogram* is a quadrilateral with two sets of parallel lines. See examples below.

Drawing a parallelogram to given dimensions. To draw a parallelogram that has sides 3" and 2", use this procedure. Start by drawing a base line slightly longer than 3". Then slide your T-square down and place your ruler under the line. Measure the 3" and draw a tick mark. See the left figure below.

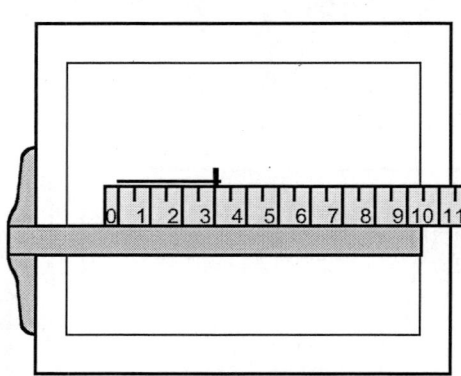
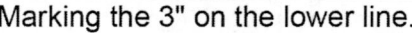
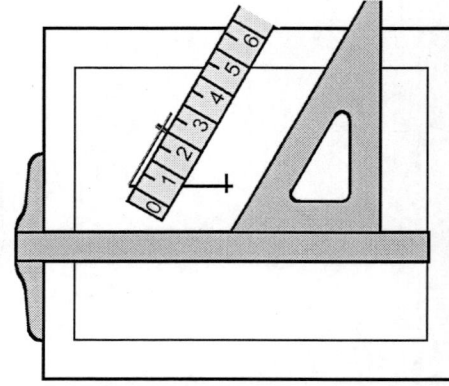

Marking the 3" on the lower line.　　　Marking the 2" on the side.

Next use your triangle to draw a side and mark 2". See the right figure above. Complete the remaining sides as shown below.

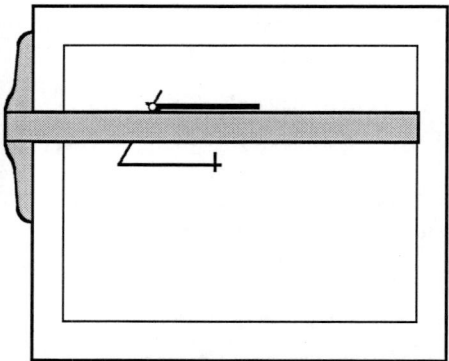
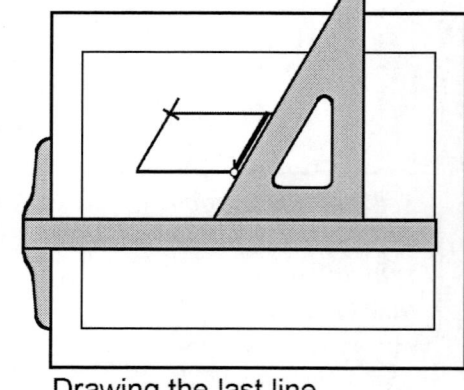

Drawing the top line.　　　Drawing the last line.

Parallelograms in the windows and bricks.

Worksheet. Draw three different parallelograms — you choose the shape — with perimeters equal to 8".

Lesson 14

Drawing Rectangles

GOALS
1. To review definitions for *rectangle, right angle,* and *perpendicular*
2. To draw rectangles with a given perimeter, given one side

MATERIALS
Worksheet 14-1 (in cm) or 14-2 (in inches)
Drawing board, T-square, 30-60 triangle
4-in-1 ruler
Two pencils or pens

ACTIVITIES
Rectangles. You probably have known what a rectangle looks like since you were in kindergarten. Now we'll think about rectangles in relationship to other polygons. First, the *rect* part in *rectangle* means *right.* So, a rectangle is a parallelogram with right angles.

Right angles. So, what is a right angle? Take two pencils and lay them flat on your table or desk. Rotate one pencil until it is straight up as shown below on the right. The two pencils form a *right angle.* In the math world the starting point measuring angles is the horizon. It also points to the right, like the 3 on a clock.

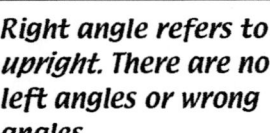

> **Right angle refers to upright. There are no left angles or wrong angles.**

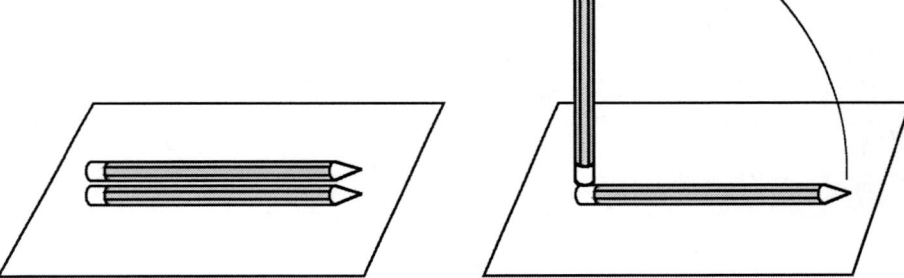

Below are more examples of right angles. How many do you see? The answer is at the bottom of the page.

When we discuss angles, the important thing is how much one line must be rotated to its new position. It makes no difference how long the lines are. Imagine rotating a toothpick or a flagpole from the horizon to the upright position. They would both make right angles with the horizon. Lines that make right angles are *perpendicular* to each other.

Worksheet. Drawing rectangles with drawing tools is easier than drawing parallelograms. Just use the right angle on your triangle. With parallelograms you had some choice on the shape of your figure. Rectangles offer no such decision making (or creativity).

For the third rectangle, visualize where the right angle will be. Then find the corresponding side of your 30-60 triangle. [Answer: 12]

This building in England has rectangular shaped windows.

Lesson 15

Drawing Rhombuses

GOALS
1. To review or learn the term *rhombus*
2. To draw rectangles with a given perimeter, given one side

MATERIALS
Worksheet 15
Drawing board, T-square, 30-60 triangle

ACTIVITIES
Rhombuses. If you looked at the figures below, you might call them diamonds. Well, the math folks have their own word, *rhombus*. Its plural is *rhombuses*, or sometimes, *rhombi*.

The House of Adam in Angers, France.

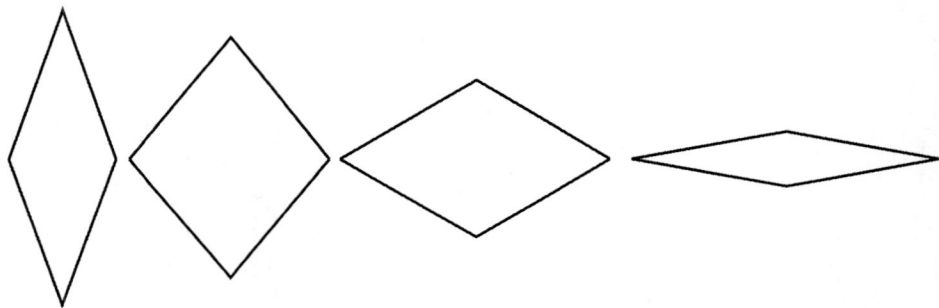

Some examples of rhombuses.

A rhombus is a parallelogram with equal sides, or you could say, an equilateral parallelogram. Rhombuses can have any orientation, as in the following examples.

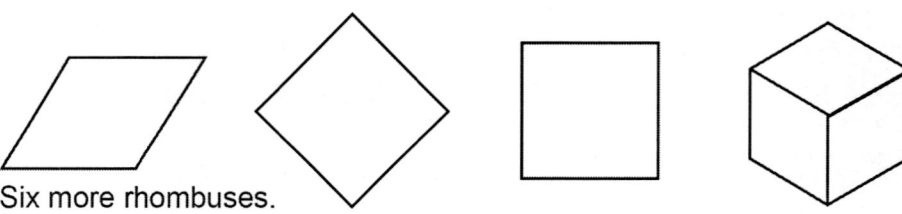

Six more rhombuses.

Are you curious why a square is here? Well, it fits the definition. A square is a parallelogram (two sets of parallel lines) with equal sides.

Worksheet. Drawing the four rhombuses on the worksheet does not require a ruler. Construct the rhombuses with your T-square and 30-60 triangle.

There is a special effect that occurs with the diagonals in a rhombus. You are to draw a parallelogram that is not a rhombus and compare the diagonals.

The Louvre (LOOV-ruh) Museum in Paris, France.

Brick design on a building.

Lesson 16

Drawing Squares

GOALS
1. To learn the meaning of *90°*
2. To learn the relationship between a square and a rectangle
3. To draw squares

MATERIALS
Worksheet 14 (previously completed)
Worksheet 16
Scissors
Two pencils or pens
Drawing board, T-square
45 triangle (not 30-60 triangle)

ACTIVITIES
90°. Take the two pencils and lay them side by side, like the 3 position on a clock. See the left figure below.

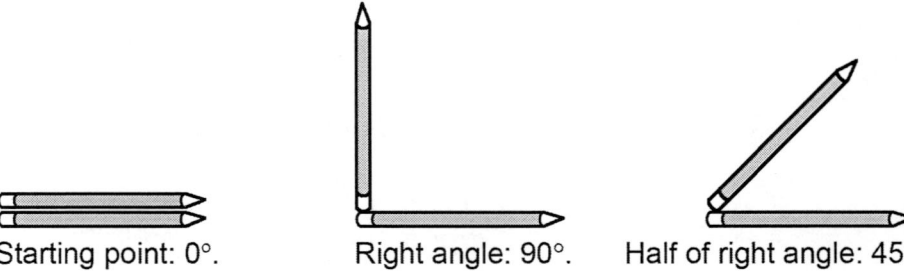

Starting point: 0°.   Right angle: 90°.   Half of right angle: 45°.

Now rotate one pencil to the 12 position; see the middle figure above. That, as you know, is a right angle, but it also called *ninety degrees*, written *90°*. The word *degree* comes from the Latin *gradus*, meaning step. Yes, the word *graduate* is also derived from *gradus*.

Now, instead of rotating to 90°, rotate half way to 90°, as shown above on the right. That is 45°, which is half of 90°. Notice the similarity to your 45 triangle.

Squares. Many people are surprised to learn that *squares* are rectangles with equal sides.

Diagonals in a square. Cut out the square on Worksheet 14, or cut out some other square. Fold it in half with two vertices touching, making a diagonal. Then compare it to your 45 triangle. See the figures below. This provides a clue to constructing a square.

Square design on a building.

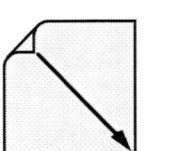

Folding a square in half, on a diagonal.

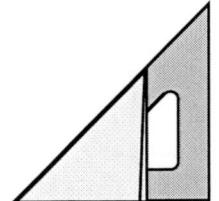

Comparing the folded square with the 45 triangle.

Worksheet. The worksheet gives directions for constructing squares. You will also draw an interesting pattern inside a square.

A common error in drawing squares is to guess the length of a side. That's a no-no.

Lesson 17

Classifying Quadrilaterals

GOALS
1. To learn the term *trapezoid*
2. To review *Venn diagrams*
3. To classify quadrilaterals

MATERIALS
Worksheets 17-1, 17-2
Drawing board, T-square, 30-60 triangle or 45 triangle

ACTIVITIES
Trapezoid. There is one more special quadrilateral, a *trapezoid*. A trapezoid has one and only one set of parallel lines. See below for several examples.

Some examples of trapezoids.

Venn diagrams. Venn diagrams are named after John Venn, who invented them in 1918. They are a visual way to show relationships. For example, in the diagram below, polygons are divided into two groups, quadrilaterals and triangles. Since the hexagon isn't part of either group, it remains outside the circles (actually ellipses).

Polygons

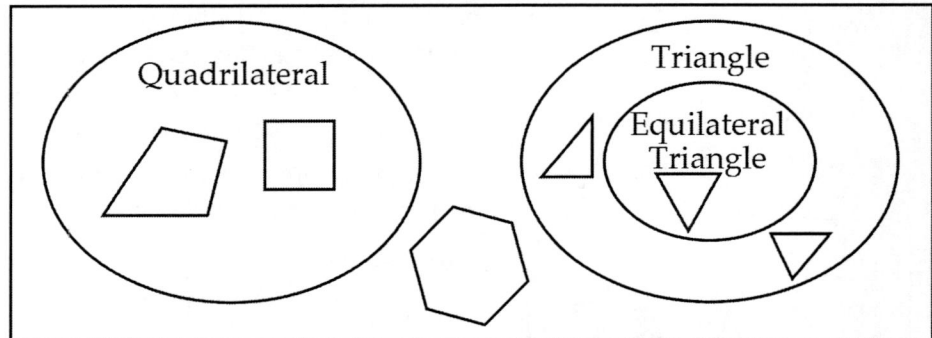

A Venn diagram.

In the quadrilateral circle are two quadrilaterals. The Triangle circle has another circle inside it. Since all equilateral triangles are triangles, the Equilateral Triangle circle is drawn completely inside the Triangle circle. Three triangles are in the appropriate circles.

Sometimes circles overlap in a Venn diagram when an item belongs to both categories.

Worksheets 17-1 and 17-2. On the first worksheet is a quadrilateral chart. Study the chart, which summarizes the shapes you have been drawing. Use it to answer the questions. You will also be asked to draw sample figures.

The second worksheet has a Venn diagram on Quadrilaterals for you to complete. There is no need to tape it to your drawing board.

Lesson 18

The Fraction Chart

GOALS
1. To construct missing lines in the fraction chart
2. To review or learn some relationships between fractions

MATERIALS
Worksheet 18
Drawing board, T-square, 30-60 triangle

ACTIVITIES

> *Originally the word fraction meant to break something into pieces, like fracturing a bone.*

The fraction chart. The fraction chart, shown at the right, has ten rows. The top row is 1, or the whole. The second row is the whole divided into two equal parts. The third row is divided into three equal parts. The pattern continues to the tenths.

Worksheet. On the worksheet, you are to construct the missing lines and write the fractions.

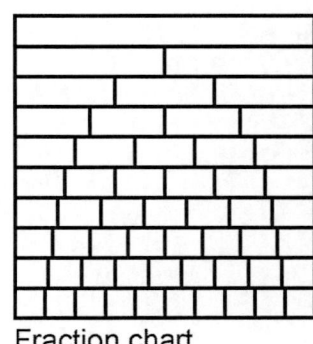
Fraction chart

The figures below demonstrate bisecting the top line and drawing the line for the half mark. You will also need the half mark for several other lines.

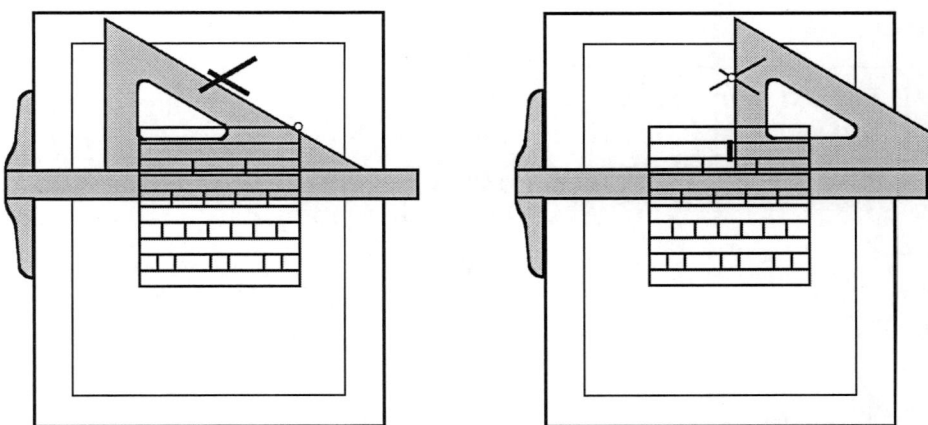

Using tick marks to determine where to draw the line for $\frac{1}{2}$.

To draw the lines for the fourths, first extend the half line to the fourths' line. Then divide each half into half again. See below. Repeat for the next fourth line. Complete the chart and answer the questions.

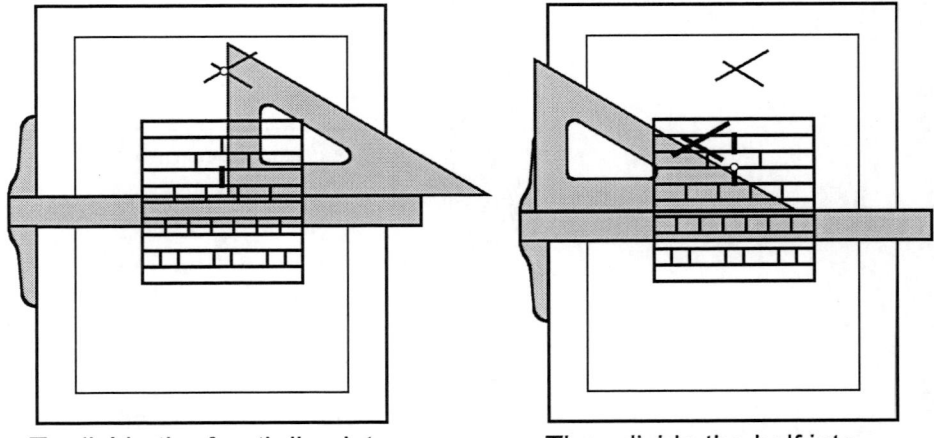

To divide the fourth line into fourths, first draw the half line.

Then divide the half into fourths using tick marks.

Lesson 19

Patterns in Fractions

GOALS
1. To review the terms *numerator* and *denominator*
2. To observe the pattern when the numerator is one and the denominator increases
3. To observe the pattern when the numerator is one less than the denominator

MATERIALS
Worksheet 19
Drawing board, T-square, 30-60 triangle

ACTIVITIES
Parts of a fraction. It is obvious that 37 is **one** number, although it is composed of **two** digits, 3 and 7. Likewise, a fraction is one number, but it has three parts. See the figure below.

The meaning of **numerator** is easy to guess. It comes from the word *number* and tells the number of parts.

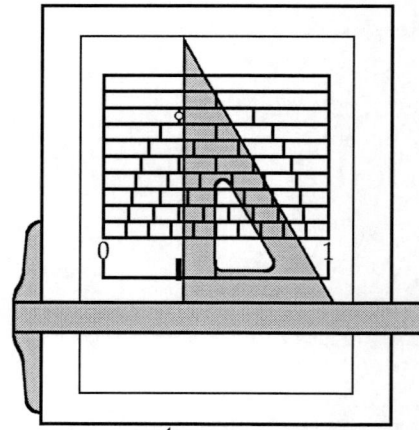

Parts of a fraction.

The word **denominator** is related to the word *nominates*, which means to name. The denominator's job is to name the size of the part.

What keeps the numerator and denominator separate is the dividing line. A fraction is really a division: the numerator divided by the denominator. In the example above, $\frac{1}{2}$ can be thought of as "one divided by two."

> **This is a short lesson. You might have time to start the next lesson.**

Worksheet. You are to draw the unit fractions on a number line. Unit fractions are fractions with a numerator of 1. Do this by projecting each unit fraction on the number line. See figures below.

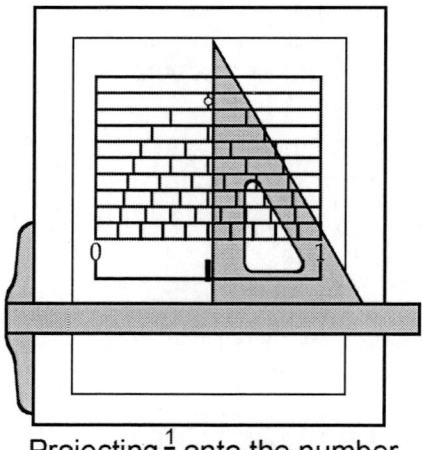

Projecting $\frac{1}{2}$ onto the number line.

Projecting $\frac{1}{3}$ onto the number line.

Line up the $\frac{1}{2}$ mark with your triangle and draw the line on the number line. See the left figure above. Drawing the $\frac{1}{3}$ isn't any harder; line up the $\frac{1}{3}$ mark and draw it on the number line. See the right figure above.

Continue in the same way to draw all the unit fractions. Then label the fractions. The worksheet also has another fraction pattern.

Lesson 20

Measuring With Sixteenths

GOALS 1. To construct a "giant" ruler with sixteenths
2. To practice reading fractions to sixteenths

MATERIALS Worksheet 20
Drawing board, T-square, 30-60 triangle

ACTIVITIES ***A ruler with sixteenths.*** Shown below is the ruler with sixteenths that you will be constructing. First you find the half by bisecting the number line. Draw the line to the height of line *a*, shown at the left.

Divide each half into halves; make them the height of line *b*. These are the fourths. Divide the fourths in half to make eighths; draw them the height of line *c*. Make the sixteenths the height of line *d*.

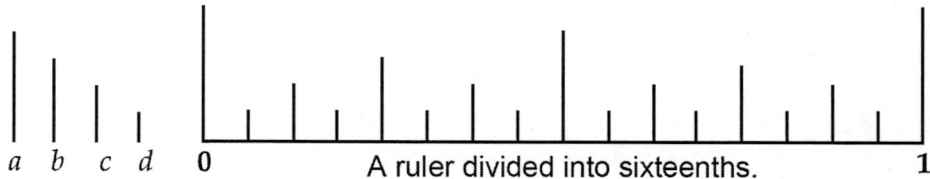

a b c d 0   A ruler divided into sixteenths.   **1**

Worksheet. On the worksheet, start by dividing the number line in half, shown by the two tick marks in the left figure below. Draw the half line above the number line. The right figure shows making a tick mark for determining the height. Repeat the process dividing the two halves into fourths, dividing the fourths into eighths, and the eighths into sixteenths.

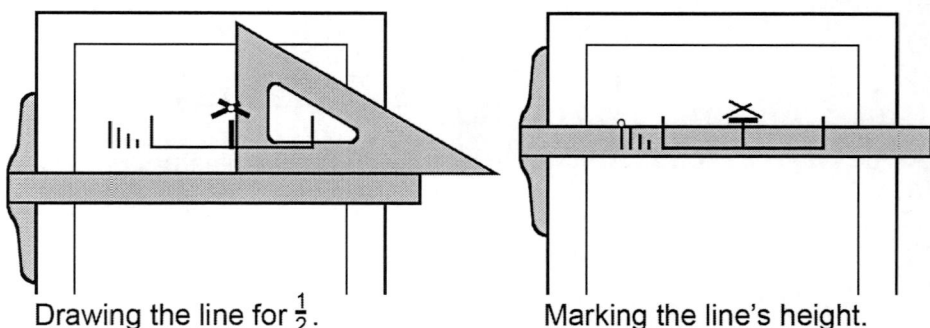

Drawing the line for $\frac{1}{2}$.   Marking the line's height.

Measuring lines. To draw a line representing a fraction, first find the fraction, $\frac{3}{4}$, in this case, and draw the tick mark. See left figure below. Next draw the horizontal line from the × to the tick mark, as shown in the right figure.

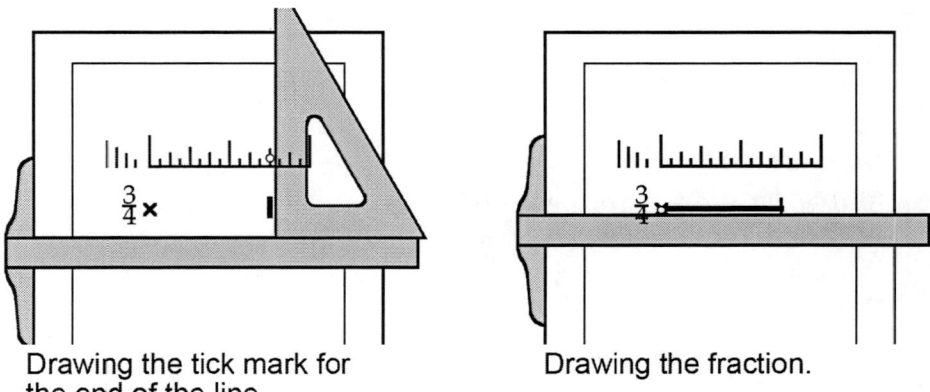

Drawing the tick mark for
the end of the line.   Drawing the fraction.

Lesson 21

A Fraction of Geometry Figures

GOALS
1. To learn to *crosshatch*
2. To find a fraction of a geometric figure

MATERIALS
Worksheet 21
Drawing board, T-square, 30-60 triangle, 45 triangle

ACTIVITIES
Crosshatching. Remember back in the lower grades when you were asked to show half of a circle? To "shade" it, you probably dug out your crayons to color the area. Or maybe you furiously scribbled in the space with your stubby pencil. Well, engineers and designers have a more sophisticated method of showing shading, called crosshatching. Computer software, called CAD for Computer Aided Design, can crosshatch automatically.

Crosshatching a circle. Let's say your task is to shade half of a circle. First find the half. Then take your 45 triangle and fill in the space with evenly spaced lines. The lines should be about this far apart (—), which is about 0.3 cm ($\frac{1}{8}$ inch). See the process in the three figures shown below.

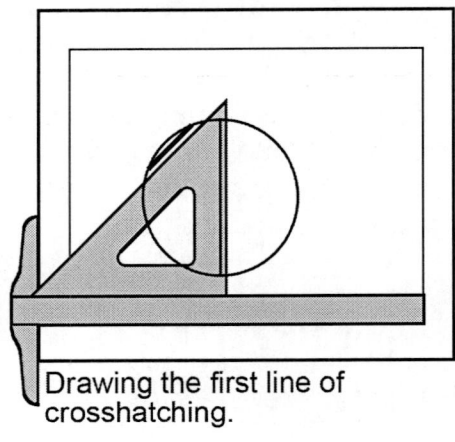

Drawing the first line of crosshatching.

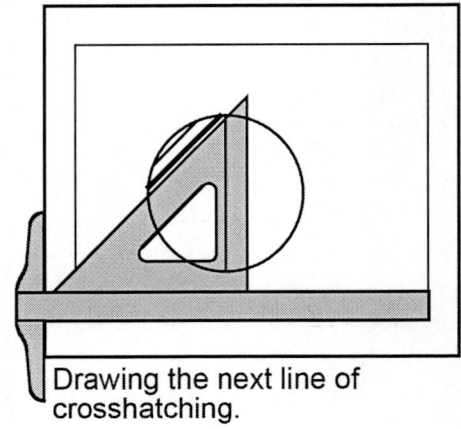

Drawing the next line of crosshatching.

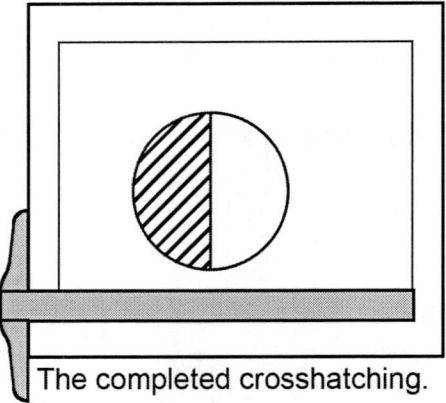

The completed crosshatching.

> There is more than one answer for these problems. For example, "C" has 12 answers.

Worksheet. For each figure, first find the requested fraction. A few of these will take some thinking. Don't be fooled by too many, not enough, or the wrong lines. Draw your own when you need to. Then crosshatch the area.

Lesson 22

Making the Whole

GOAL   1. To construct the whole, or 1, given the fraction

MATERIALS   Worksheet 22
Drawing board, T-square, 30-60 triangle, 45 triangle

ACTIVITIES   ***Worksheet.*** This worksheet doesn't take much explanation, but it will take some thinking on your part. In each rectangle are a figure and a fraction. You are to complete the figure so it is a whole. For example, problem A shows a triangle and the fraction, $\frac{1}{2}$.

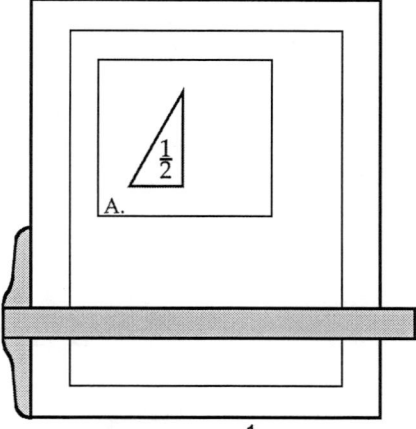

A triangle that is $\frac{1}{2}$ the figure.

One solution is to construct another triangle as shown below.

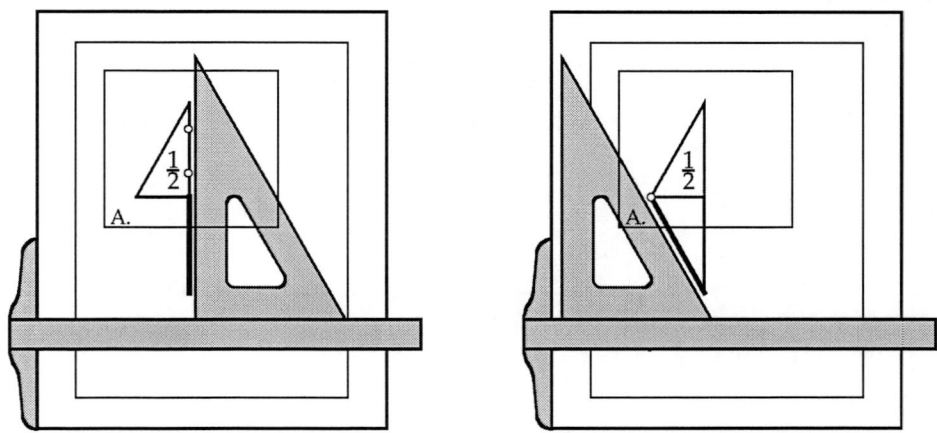

Constructing another triangle equal to the original triangle.

The only problem with this solution is that it doesn't fit within the rectangle. There are three other solutions that will fit. Choose one of them to construct. Then complete the remaining figures.

Lesson 23

Ratios and Nested Squares

GOALS
1. To learn the term *ratio*
2. To learn to write ratios
3. To construct a series of nested squares

MATERIALS
Worksheet 23
Drawing board, T-square, 45 triangle
Scissors

ACTIVITIES
Ratios. Whether you've heard of ratios or not, they're not very hard. A *ratio* (RAY-shee-o) is two numbers telling how many times greater or smaller one object is compared to another. There are three ways to make ratios and three ways to write a ratio.

An example is the best way to understand ratios. In the figure below, the ratio of the area of the black triangles to the large triangle is 3 to 4. That was a *part to whole* ratio.

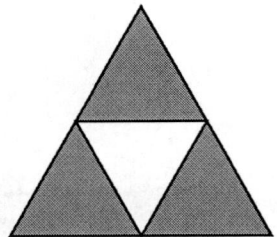

The ratio of the large triangle to the small white triangle is 4:1, read as 4 to 1. That is a *whole/part* ratio, the second type of ratio.

The ratio of the small white triangle to the black triangles is $\frac{1}{3}$. That, of course, is a *part/part* ratio, the third type of ratio.

All this is summarized in the chart below.

Type	Triangles (area)	Ways to write ratios		
Part to whole	Black to large	3 to 4	3:4	$\frac{3}{4}$
Whole to part	Large to white	4 to 1	4:1	$\frac{4}{1}$
Part to part	White to black	1 to 3	1:3	$\frac{1}{3}$

Types of ratios and ways to write them, using the figure above as the example.

Two squares. Start Worksheet 23 halfway down, instead of the top. This strange beginning allows you to cut up part of the page while leaving enough paper to draw the day's masterpiece. Draw the square. Refer to Worksheet 16 if you need a refresher.

To draw the smaller square, first find the larger square's center. This is shown on the left at the top of the next page. Then you can find the midpoint of the top line. See the figure on the right. The rest is easy as the next two figures show.

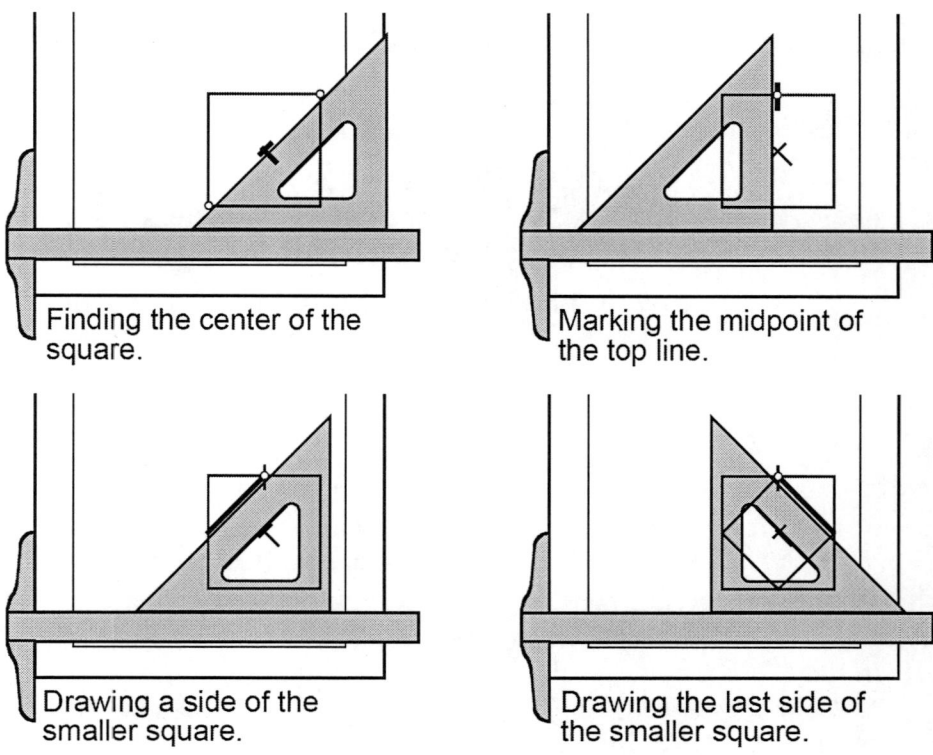

Finding the center of the square.

Marking the midpoint of the top line.

Drawing a side of the smaller square.

Drawing the last side of the smaller square.

Cut out the larger square and fold on the lines of the smaller square. What do you notice? What is the ratio of the area of the smaller square to the larger square? Answers are at the bottom of the page.

Now cut out the smaller square, saving the triangles. Arrange all four triangles into one square. How does it compare to the smaller square? Finally, take two of the triangles and form a square. What is the ratio of the smallest square to the largest square? The answers are at the bottom of the page.

Nested squares. Complete the worksheet. To draw the third square, find the midpoint of the side of the second square. It lies on the first square's diagonal. See the left figure below. Draw the first side as shown below in the right figure. Finish the third square and draw as many additional squares as you can. [Answers: triangles cover small square, ratio 1 to 2, squares are congruent, ratio 1 to 4]

Puzzle: Give the five pieces to someone else. Ask her to use all the pieces to make a) one square or b) two squares or c) three squares.

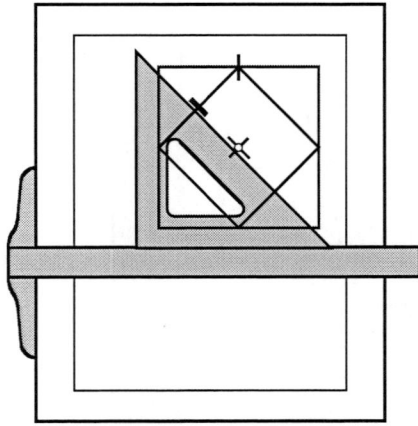

Finding the midpoint needed for the third square.

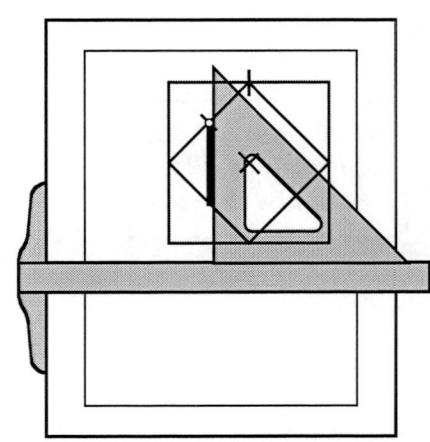

Drawing the first side of the third square.

28

Lesson 24

Square Centimeters

GOALS   1. To understand the term *square centimeter*
2. To define *area*

MATERIALS   Worksheet 24
About fifteen centimeter cubes, optional
Drawing board, T-square, 30-60 triangle, 4-in-1 ruler

ACTIVITIES   ***Measuring area.*** Guess which rectangle below needs more little squares to completely fill the area inside it. Answer below.

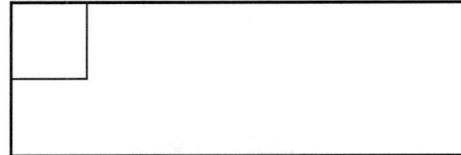
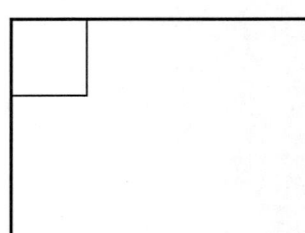

Which rectangle has more area?

> In the 1790s, French scientists developed the metric system. In 1840, France converted to it use. In the 1870s, scientists adopted it.
>
> United Kingdom began converting in 1965; Australia, in 1970; and Canada in 1975.

In the metric system, we measure lengths in centimeters, meters, kilometers (KIL-oh-mee-ter), and so forth.

To measure area, we usually use square units. In the metric system the sides of the squares could be 1 cm, 1 m (meter), or 1 km (kilometer). We call these unit squares: square centimeter, square meter, and square kilometer. **Area** is the number of such squares it takes to cover a surface.

If you have never done so before, cover the rectangles above with square centimeters found on the centimeter cubes. This will help you understand the concept of area.

Worksheet. Answer the first question by making a guess. Next you are to fill in the area with square centimeters.

> Be sure to use the centimeter scale on your ruler.

To draw the square cm, use your ruler and mark off cm marks along one side and either the top or bottom as shown below on the left. Draw these tick marks freehand, that is, without the T-square or triangle. Next construct the square cm by drawing the horizontal and vertical lines. See the right figure below. Find the number of square cm in each rectangle. [Answer: the same]

> Notice how the square centimeters fill the space like the cubes.

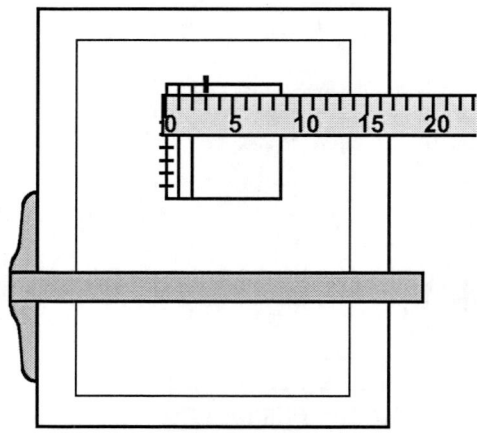

Marking the centimeters.

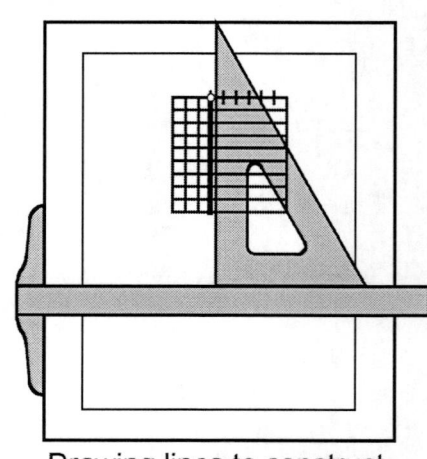

Drawing lines to construct square centimeters.

G: © Activities for Learning, Inc. 2010

Lesson 25, optional

Square Inches

GOALS
1. To understand the term *square inch*
2. To define *area*

MATERIALS
Worksheet 25
About twenty 1" tiles, optional
Drawing board, T-square, 30-60 triangle, 4-in-1 ruler

ACTIVITIES
Measuring area. Guess which rectangle below has the greater area. The answer is at the bottom of the page.

Which rectangle has more area?

In 1866 the U.S. Congress legalized the use of the metric system.

Pharmacists and the U.S. Army and Marine Corps converted in the 1950s. NASA switched in the 1960s.

All automobile measurements are metric.

In the U.S. Customary and British imperial systems, the units of length are inches, feet, yards, miles, and so on. These units have been used for over a thousand years. In contrast, the metric system is only about 200 years old.

To find areas, we usually use squares of these units, square inches, square feet, square yards, and square miles. **Area** is defined by the number of such squares it takes to cover a surface.

Worksheet. Answer the first question by making a guess. Next you are to fill in the area with square inches.

To draw the square inches, use your ruler and mark off inch marks along one side and either the top or bottom as shown below on the left. Draw these tick marks freehand. Next construct the square inches by drawing the horizontal and vertical lines. See the right figure below. Find the number of square inches in each rectangle.
[Answer: right rectangle]

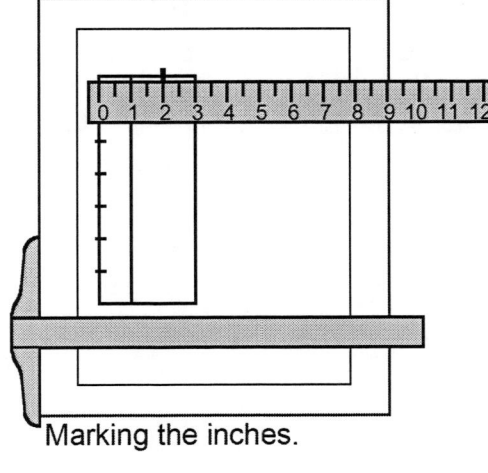

Marking the inches.

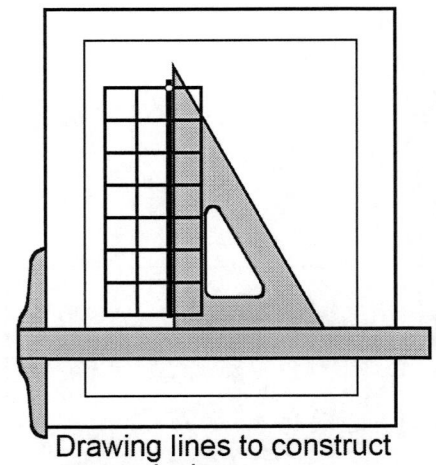

Drawing lines to construct square inches.

Lesson 26

Area of a Rectangle

GOALS
1. To understand how area is calculated
2. To see the connection between multiplication and area
3. To learn and apply the *formula* for the area of a rectangle

MATERIALS
Worksheet 26
Drawing board, T-square, 30-60 triangle

ACTIVITIES
Worksheet. Begin by completing Questions 1-4 on the worksheet.

Formula for finding the area of a rectangle. A formula is a general principle stated in mathematical symbols. The word *formula* means "little form." So, it is a shortcut for stating a mathematical relationship. Most of the time you do not need to just memorize formulas. Rather, they are a logical result you can think through.

In question 5 on the worksheet, you were asked to write the formula for the area of a rectangle. The number of squares needed to cover a figure is the area, usually written as *A*. You found the area of all those rectangles by multiplying the number of squares in a row by the number of rows. So, if we call the horizontal distance, the width *w*, and the vertical distance, the height *h*, the formula becomes

$$A = w \times h$$
$$A = wh$$

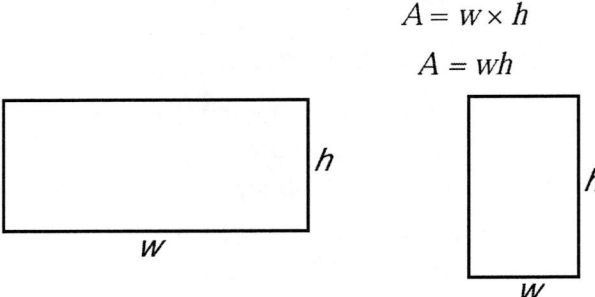

> **In some math textbooks (especially older texts), the symbol "l" for length is used rather than "h."**
>
> **Today, computer software usually uses width and height.**

Actually, in algebra, which has lots of formulas, the operator "×" is not written. Two letters written together without an operator means multiply. This is one of the major differences between arithmetic and algebra. In arithmetic, digits written side by side without an operator mean add; for example, 976 is 900 + 70 + 6 and $3\frac{1}{2}$ means $3 + \frac{1}{2}$. However, computer spreadsheets require operators between all numbers and letters like *w* and *h*.

Symbol, cm², for square centimeter. The symbol, cm^2, is the abbreviation for square centimeter. Just as $3^2 = 9$ and forms a square with 3 on a side, so $1\ cm^2$ is a square that is 1 cm on a side. Read $1\ cm^2$ as "1 centimeter squared."

Problem 5. How would you find the area of the figure below? Think about several ways and then discuss it with a partner, if possible, before looking at the solutions on the next page.

A building in England with many rectangles.

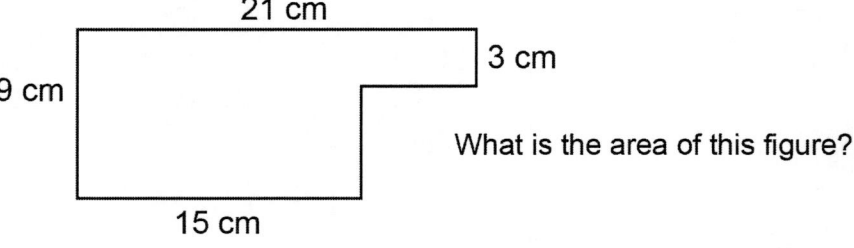

What is the area of this figure?

G: © Activities for Learning, Inc. 2010

There are several ways to solve this problem.

Solution 1. Make the whole figure into a rectangle and then subtract the little rectangle.

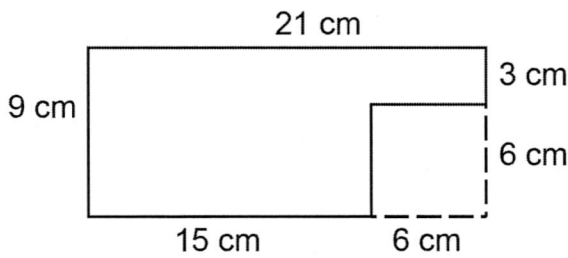

$A = wh$ (large rectangle) $- wh$ (square)

$A = 21 \times 9 - 6 \times 6$

$A = 189 - 36$

$A = 153$ cm^2

Solution 2. Divide the rectangle horizontally into two rectangles.

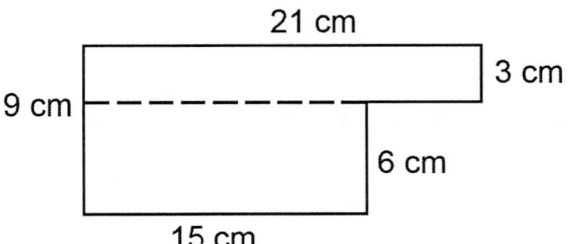

$A = wh$ (upper rectangle) $+ wh$ (lower rectangle)

$A = 21 \times 3 + 15 \times 6$

$A = 63 + 90$

$A = 153$ cm^2

Solution 3. Divide the rectangle vertically into two rectangles.

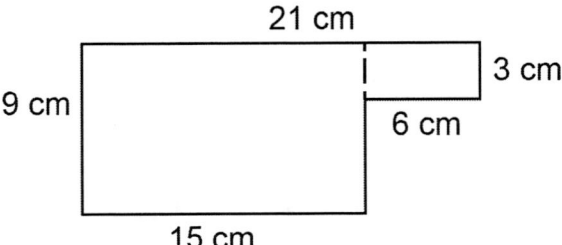

$A = wh$ (left rectangle) $+ wh$ (right rectangle)

$A = 15 \times 9 + 6 \times 3$

$A = 135 + 18$

$A = 153$ cm^2

Lesson 27

Comparing Areas of Rectangles

GOALS
1. To calculate more areas of rectangles
2. To compare areas of rectangles with constant perimeter

MATERIALS
Worksheets 27-1, 27-2
Drawing board, T-square, 30-60 triangle
4-in-1 ruler

ACTIVITIES
Frame problem. Consider the following problem. You have 12 cm of gold edging to place around a rectangular frame. You want the maximum amount of space inside the frame.

First think about the possible dimensions of the rectangles, so the perimeters will be 12 cm. Then study the figures below.

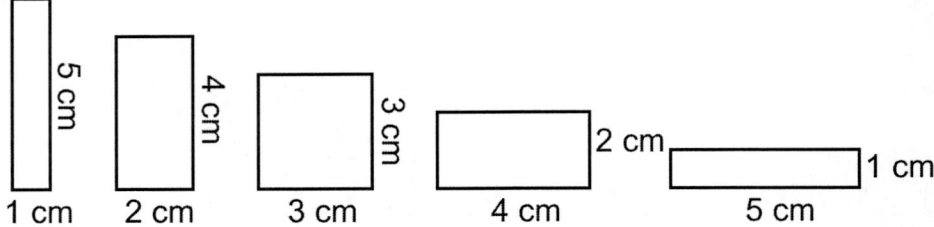

The areas, which you can do in your head using $A = wh$, are from left to right, 5 cm^2, 8 cm^2, 9 cm^2, 8 cm^2, and 5 cm^2.

Graphing the frame problem. It is interesting to graph the results as shown below. Why is the area equal to 0 when the width is equal to 0 or 6?

> *This type of problem is easily solved with a branch of mathematics called calculus.*

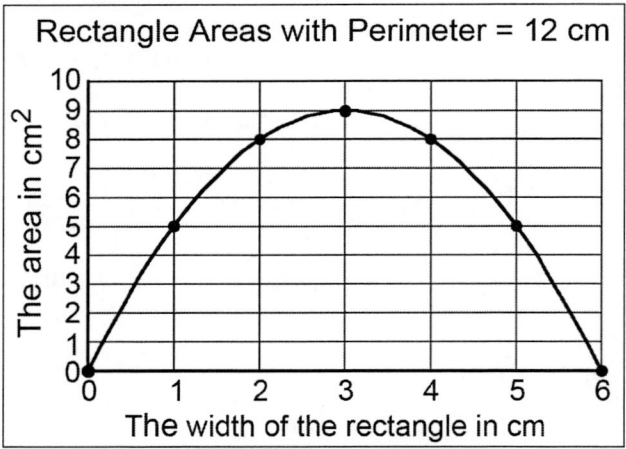

The shape of this graph is called a *parabola.*

You can see the greatest area occurs when the width of the rectangle is to 3. What is the height when the width is 3? The answer is at the bottom of the page.

Worksheets. There is a similar problem on Worksheets 27-1 and 27-2. Draw the rectangles by measuring with your ruler like you did on Worksheet 11. [Answer: 3]

Lesson 28

Product of a Number and Two More

GOALS 1. To review the meaning of *exponent*
2. To investigate the relationship of a number times the number plus 2

MATERIALS Worksheet 28
Drawing board, T-square, 30-60 triangle or 45 triangle

ACTIVITIES ***Doubles in addition.*** When you were learning your addition facts, you probably learned about doubles and near doubles. If you knew, for example, that 8 + 8 = 16, then 8 + 9 was 1 more, or 17.

Similarly, there is a simple way to add a number plus 2 more than the number. Find the "middle" of the numbers and double it. For example, to add 19 + 21, the middle is 20 and 20 doubled is 40. So, 19 + 21 = 40.

Multiplying a number by the number plus 2. There is also an interesting relationship between numbers like 2×4 and 3×3. To multiply a number by itself is called squaring the number and it is written as

$$3^2 = 9$$

> *Why do you think 3×3 is referred to as "3 squared"?*

The little 2, called the *exponent*, tells how many times the 3 is multiplied by itself. For example, $4^5 = 4 \times 4 \times 4 \times 4 \times 4$.

To investigate the relationship of 2×4 and 3×3, first you need to draw a rectangle that is 2 by 4. See the left figure below.

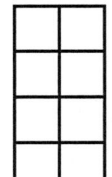 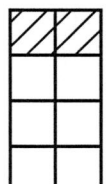 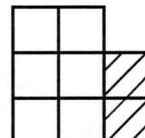 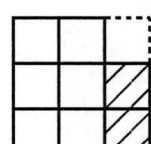

Next you need to transform the rectangle into a square. First crosshatch the top row, as shown above in the second figure. This indicates it will be moved to a new position in the next drawing.

Redraw the crosshatched squares tacked on the right side of the new figure. See the third figure above. Notice that you're close to a square. But you're one little square short, so draw it in with dotted lines. See the last figure.

Worksheet. There are three rectangles on the worksheet for you to transform into near squares.

Figure out the relationship, fill in the table, and answer the questions.

For Question 7, if *n* is the middle number, then $n - 1$ is the lower number. What is the other number? What happens when you multiply them?

Lesson 29

Area of Consecutive Squares

GOALS   1. To compare the area of consecutive squares
2. To see this relationship on the multiplication table

MATERIALS   Worksheets 29-1 and 29-2
Drawing board, T-square, 45 triangle

ACTIVITIES   ***The next square.*** This lesson compares the size of the new square formed by adding one to the sides of the original square. (*Consecutive* means the next one in order.)

Drawing the next larger square. Start by drawing an extra row of squares at the top. This means you extend the lines as shown in the left figure below. A diagonal will help you find the top of the new little squares. Draw the top line and crosshatch the new little squares. It should look like the figure on the right.

> **You can draw the diagonal, rather than a tick mark if you want. You will need to crosshatch it anyway.**

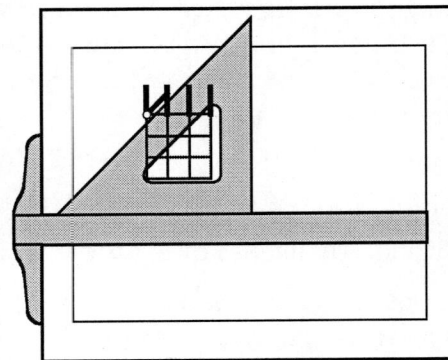 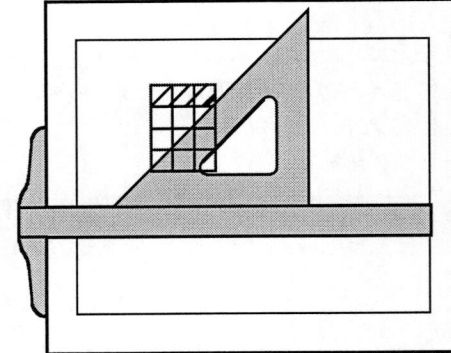

It doesn't look like a square. So you need to add a column on the right as shown below. Crosshatch the new column to make them match the new row. Draw the remaining square.

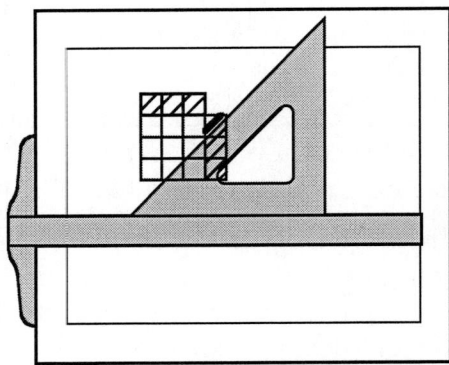 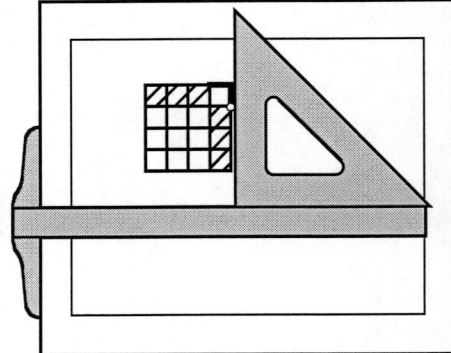

Worksheet 29-1. On this worksheet are three rectangles for you to build on to make the next larger square. Once you figure out the relationship, fill in the table, and answer the questions.

Worksheet 29-2. This worksheet does not need drawing tools. It combines concepts from several lessons.

Lesson 30

Perimeter Formula for Rectangles

GOALS 1. To learn several formulas for the perimeter of rectangles
2. To apply a perimeter formula
3. To review *factors*

MATERIALS Worksheet 30
Drawing board, T-square, 4-in-1 ruler, 30-60 triangle or 45 triangle
Basic calculator with memory

ACTIVITIES ***Formula for perimeter of a rectangle.*** You have been finding perimeters for quite a while, but we haven't discussed formulas.

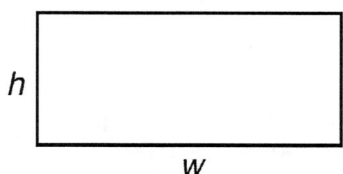

A rectangle with dimensions.

> In older textbooks, you could easily tell an area problem from a perimeter problem. The number of dimensions told you; two for area and four for perimeter.

In the rectangle above, you could find the perimeter three different ways. In the first way, you take a walk around the rectangle and write down the measurements as you go. So the formula is

$$P = w + h + w + h$$

While it's straightforward, it requires adding four numbers, two of which are the same.

The second way is to notice that you have two w's and two h's. So you can write the formula as

$$P = 2w + 2h$$

That takes two multiplications and one addition.

The third way is to add the w and the h and then multiply by 2. That formula looks like.

$$P = 2 \times (w + h)$$

To make it look like algebra, omit the "×," and write it as

$$P = 2(w + h)$$

Windows from Pike Place Market, Seattle, WA.

Perimeter on a calculator. Check which formula is most efficient using a simple calculator with memory. Using w = 3 and h = 4, count the number of keystrokes it takes to find the perimeter with each formula. The answers are at the bottom of the page.

Check which formula is most efficient using a simple calculator with memory. Using w = 3 and h = 4, count the number of keystrokes it takes to find the perimeter with each formula. The answers are at the bottom of the page.

Now try with three digits such as $w = 111$ and $h = 333$.

Worksheet. First draw as many rectangles as you can with area equal to 24 cm². Then write a perimeter formula and calculate the perimeters for each rectangle. [Answers for simple calculator: 1 digit: formula 1, 8 strokes; formula 2, 9 strokes; formula 3, 6 strokes. 3 digits: formula 1, 15 strokes; formula 2, 13.]

Lesson 31

Area of a Parallelogram

GOAL   1. To find the area of a parallelogram

MATERIALS   Worksheet 31
Drawing board, T-square, 45 triangle, 4-in-1 ruler

ACTIVITIES   ***Worksheet.*** Do Problems 1-3 before reading any farther. Then continue reading.

Area of a parallelogram. Finding the area of parallelogram is simple once you transform the parallelogram into a rectangle. That is what the first few questions on the worksheet demonstrated. This means you can use the same formula once you realize what the height h is.

The word "height," as you'd expect, is the perpendicular distance from the base of the parallelogram to the top. That's also true in a rectangle because the side is the height.

A parallelogram area problem is demonstrated for you. You can draw the height at a vertex, shown by the solid line, or some other place, shown by the dotted line.

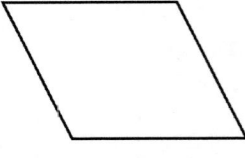
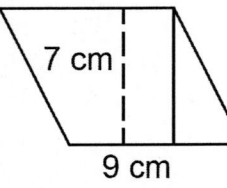

$A = wh$
$A = 9\ cm \times 7\ cm$
$A = 63\ cm^2$

Sometimes, you need to extend a side of the parallelogram in order to draw the height, as in the figure below.

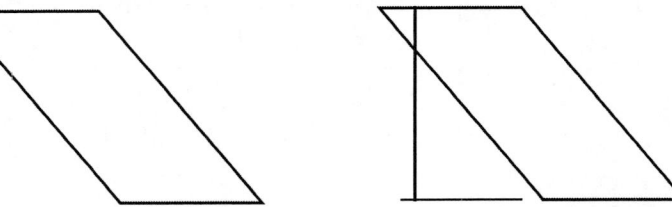

More windows at Pike Place Market in Seattle, WA.

Of course, you could extend the top edge if you prefer.

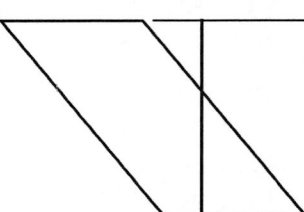

Worksheet. Complete the worksheet.

Lesson 32

Comparing Calculated Areas of Parallelograms

GOALS
1. To review *millimeters* and *square millimeters* (■)
2. To calculate the area of parallelograms in mm²
3. To compare areas calculated different ways

MATERIALS
Worksheet 32
Drawing board, T-square, 45 triangle, 4-in-1 ruler
Calculator, optional

ACTIVITIES
Measuring in millimeters. Since there are 10 millimeters in a centimeter, you can use the centimeter scale to find millimeters. For example, 2 cm is 20 mm. Each of the tenths of a centimeter is 1 mm.

> The abbreviation of millimeter is "mm" without a period.

Finding the height when the width is not horizontal. That's a long title. It describes the complication when the parallelogram looks like the figure on the right.

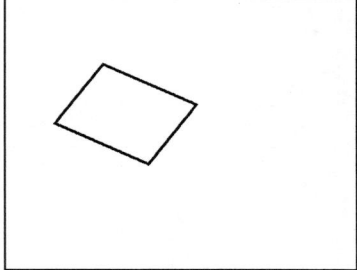

The parallelogram.

Turn your T-square upside (it will then lie flatter) and place the 45 triangle on it. Move the two pieces as a unit and align the triangle along the side of the parallelogram that you consider to be the width. See the second figure.

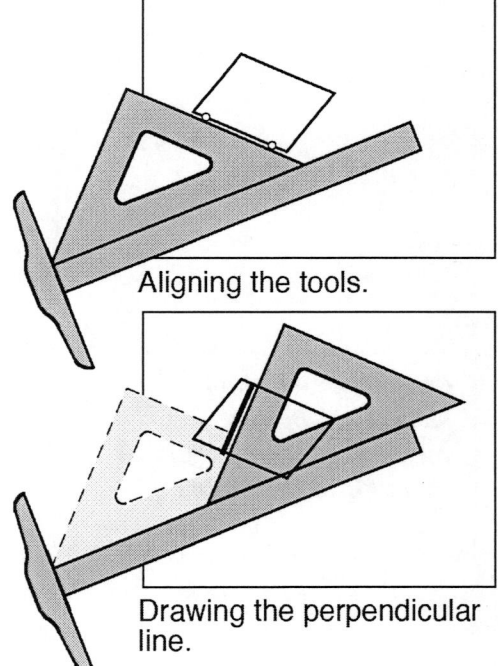

Aligning the tools.

Keep the T-square from moving and slide the triangle so you can draw the perpendicular line where you want it. If your T-square moves, you must start over. Than draw the perpendicular line as shown.

Drawing the perpendicular line.

Worksheet. Complete the worksheet.

Comparing results. After you have calculated the areas two different ways, you will notice that your answers are not exactly the same. This occurs because your measurements are not accurate enough. If you measured to the tenth of a millimeter, your answers would be closer.

Lesson 33

Area of a Triangle

GOALS
1. To compare a triangle to a parallelogram
2. To find the area of triangle

MATERIALS
Worksheet 33
Drawing board, T-square, 45 triangle, 4-in-1 ruler

ACTIVITIES
Diagonals in a parallelogram. Look at the congruent parallelograms below. Each is divided into two parts by a diagonal. Does it seem to you that the diagonals cut the parallelogram into two congruent parts? To be certain, cut out the triangle (Worksheet 33 has a larger version) formed by two sides and the diagonal and see if it matches the remaining part.

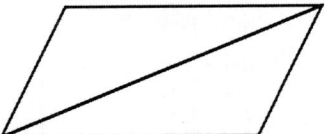 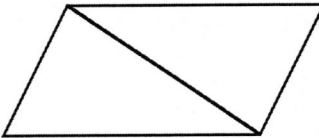

Area of a triangle. Okay, now you're convinced, so we can proceed. If a diagonal divides a parallelogram into two congruent triangles, then a triangle must be half of a parallelogram. So, if you enclose the triangle in a parallelogram, you could find the area of the triangle because it is half the area of the parallelogram. Below are some examples of enclosing parallelograms around a triangle.

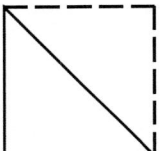 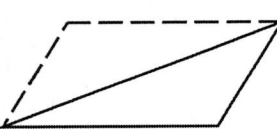 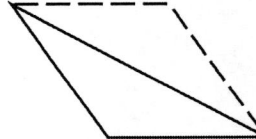

Formula for area of a triangle. If the area of a parallelogram is

$$A = wh,$$

then the area of a triangle must be

$$A = \frac{1}{2}wh \text{ or } A = \frac{wh}{2}$$

> *Remember, formulas should not be rote memorized. Think them through so they make sense.*

Worksheet. First you are to draw the enclosing parallelogram. Except for rectangles, this takes a technique similar to the one in the previous lesson. See the figures below.

> *Be sure to keep the T-square from moving.*

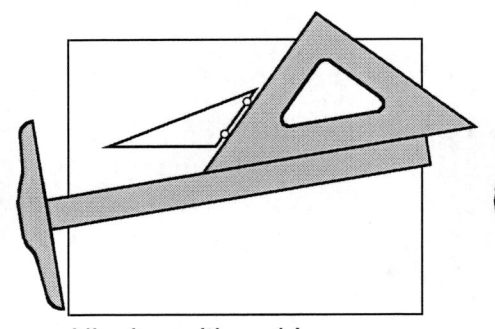 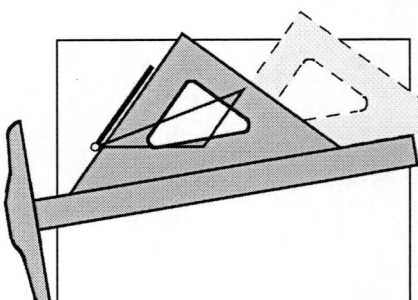

Aligning with a side.   Drawing the parallel line.

Lesson 34

Comparing Calculated Areas of Triangles

GOALS
1. To compare areas calculated three different ways
2. To learn the meaning of the *little square* at the intersection of two lines
3. To calculate the area of triangles to the nearest tenth of a cm²
4. To learn the term *altitude*

MATERIALS
Worksheet 34-1 (in cm) or 34-2 (in inches)
Drawing board, T-square, 30-60 triangle, 4-in-1 ruler
Calculator

ACTIVITIES
Finding areas of triangles in different ways. In the previous lesson, when you found the area of a triangle, you probably used the horizontal line as the width. You know in parallelograms that the width can be any side. Well, it's also true for triangles; the width can be any side.

Below is the same triangle, each with a different side chosen to be the width.

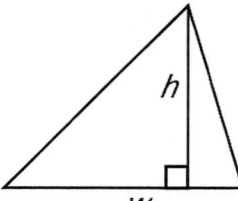 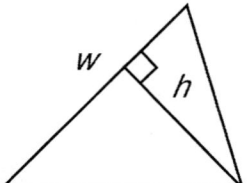 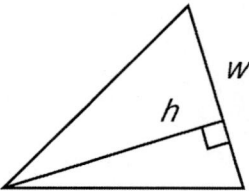

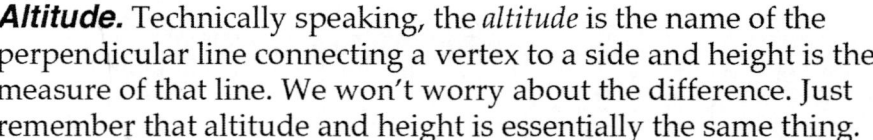
The little square at the intersection of two lines means the lines are perpendicular.

Altitude. Technically speaking, the *altitude* is the name of the perpendicular line connecting a vertex to a side and height is the measure of that line. We won't worry about the difference. Just remember that altitude and height is essentially the same thing.

Worksheet. Do you think the area would be same if you used a different side for the width? Has the area changed because you used a different side?

This is the problem you will investigate on the worksheet. There is one triangle for you to calculate the area three different ways by using different sides for the width. Of course, you must also draw the correct perpendicular line to match the width.

Note that the instructions ask you to calculate the answer to the nearest tenth. That means, for example, if your answer is 41.86, you need to round it to 41.9.

Answers agreeing. After you find the areas by different methods, you will probably notice that your answers are not exactly the same. The reason is that you cannot measure exactly. Rounding can also cause errors. These are problems that frequently occur in the real world.

40

Lesson 35

Converting Inches to Centimeters

GOALS
1. To compare inches and centimeters
2. To memorize the conversion 1 inch = 2.54 centimeters

MATERIALS
Worksheet 35
Either triangle, 4-in-1 ruler
Calculator

ACTIVITIES
Inches and centimeters. Some people know the length of an inch because they have a knuckle that measures 1 inch. Likewise, they remember a centimeter because their little finger or fingernail measures 1 cm. Check what works for you. Is a Popsicle stick 1 inch or 1 cm wide? Answer is below.

Tenths of an inch. Most rulers with inches split the inch into halves, fourths, eighths, sixteenths, and so forth. If the ruler designer wanted even smaller divisions, what would the next unit be? The answer is at the bottom of the page.

To find perimeter in inches using sixteenths of an inch may require adding fractions with unlike denominators. We will avoid this by using tenths of an inch.

On your 4-in-1 ruler is an unusual scale — tenths of a inch, which is shown below. Below the inch scale is a centimeter scale.

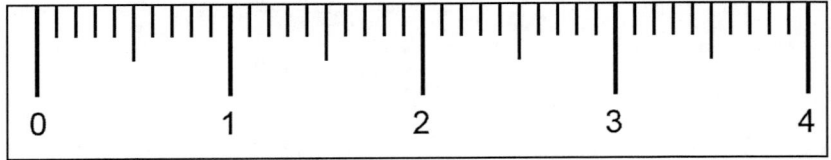

The upper ruler shows inches divided into tenths of an inch. The lower ruler shows centimeters divided into tenths of a cm.

One way to remember "1 in. = 2.54 cm" is $1 is 25 (quarter) × 4.

Use your triangle as shown on the right and estimate how many centimeters are in a inch. The usual conversion is 1 inch = 2.54 cm. So, how centimeters are in 2 in? How many centimeters are in 3.4 in.? Answers are below.

Worksheet. The worksheet has two triangles. You are to calculate the perimeters both to the nearest tenth of an inch and the nearest tenth of a centimeter. Record your results in the chart at the bottom of the page. Then you are to calculate the number of centimeters by applying the conversion to the perimeter in inches.
[Answers: 1 cm; thirty-seconds; 2 × 2.54 = 5.08 cm; 8.64 cm]

Finding the number of centimeters in an inch.

Lesson 36

Name That Figure

GOALS 1. To practice identifying figures with letters
2. To review the term *isosceles*

MATERIALS Worksheet 36

ACTIVITIES ***Naming figures.*** When we want to name a certain polygon among a crowd of others in a figure, we use letters to name the vertices. Usually, we use italic capital letters.

For example, in the figure below *ABF* names a small triangle while *AEC* names a large triangle. *FDE, FED, EFD, EDF, DEF*, and *DFE* all name the same triangle. Would that work with your initials?

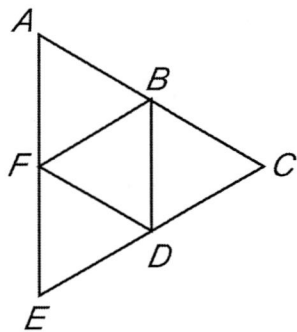

Things are a little more complicated when naming a polygon with more sides. The trapezoid *ACDF* does not need the letter *B* because *B* is not a vertex in the trapezoid. You can also name it *CDFA* or *FDCA*, but not *FCDA*. In other words, you can name the trapezoid by going around in either direction, but skipping around is not allowed.

Isosceles triangles. The definition of *isosceles* triangle is simple; it's a triangle with two equal sides. The tough part is pronouncing the word, (i-SOS-sah-leez) and spelling it. The figure below might help you to spell it.

i 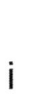 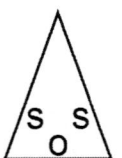 c 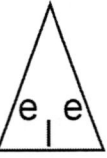 s

A spelling aid for isosceles.

An *isosceles* trapezoid is a trapezoid with the two non-parallel sides equal. See the example below.

This is an isosceles trapezoid.

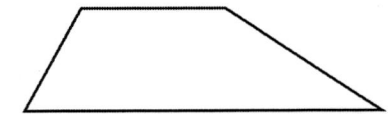

This is NOT an isosceles trapezoid.

The Louvre Museum in Paris, France.

Worksheet. On the worksheet is a hexagon with the vertices named. You will be naming all kinds of figures.

Lesson 37

Finding the Areas of More Triangles

GOALS 1. To practice finding areas of triangles
2. To discover some interesting area results

MATERIALS Worksheets 37-1, 37-2
T-square, 45 triangle, 4-in-1 ruler
Calculator

ACTIVITIES ***Naming triangles.*** Naming vertexes with letters often makes it easier to explain to others (and yourself) what you're talking about.

For example, in the figure below, using letters makes it much simpler to explain which triangle you are discussing. Let's try saying in symbols that the area of the upper left triangle is one-half the area of the square. It looks like this

$$A(\triangle ABC) = \tfrac{1}{2} A(ABCD),$$

where A means area and $\triangle$ means triangle.

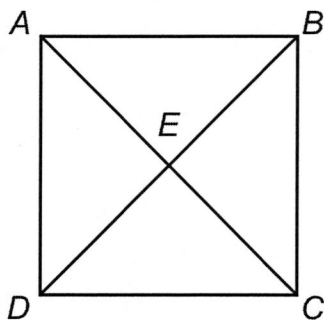

Worksheets. Both the "kite" problem and the flag problem on Worksheet 37-1 have interesting answers.

Worksheet 37-2 has two identical quadrilaterals. You are to find the area in two different ways. If you need to, refer to Lesson 32 on how to use your drawing tools to draw perpendicular lines.

Glass ceiling in England.

Lesson 38   **Area of Trapezoids**

GOALS   1. To understand the formula for the area of a trapezoid
2. To review the *distributive property*
3. To learn the term *straightedge*

MATERIALS   Worksheet 38-1, 38-2
Drawing board, T-square, either triangle

ACTIVITIES   ***Distributive property.*** The distributive property is one of the basic principles in algebra. Don't worry; it's not hard to understand.

> **The distributive property is sometimes called the "distributive law."**

For example,

$$(\tfrac{1}{2} \times 8) + (\tfrac{1}{2} \times 4) = \tfrac{1}{2} \times (8 + 4)$$
$$4 \quad + \quad 2 \quad = \quad 6$$

This really isn't startling news. Since $\tfrac{1}{2} \times 8$ also means 8 halves, then the equation becomes 8 halves + 4 halves = 12 halves. So the distributive property just means that if you are multiplying (or dividing) a bunch of numbers by the **same** number, you can add first and do just one multiplication.

Now we'll apply this to trapezoids.

Area of a trapezoid. To find the area of the trapezoid, drawn below on the left, one way is to divide it into two triangles, as shown below on the right.

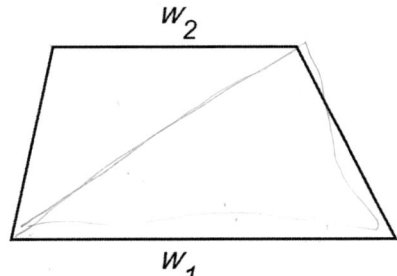   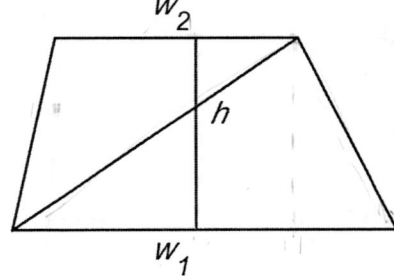

The area of the two triangles is

$$A = \tfrac{1}{2} w_1 h + \tfrac{1}{2} w_2 h$$

Apply the distribute property with the $\tfrac{1}{2}$

$$A\,(\text{trap}) = \tfrac{1}{2}(w_1 h + w_2 h)$$

Apply the distribute property again, this time, with the h

$$A\,(\text{trap}) = \tfrac{1}{2}(w_1 + w_2)h$$

Worksheet. The worksheet has another way for thinking of the area of a trapezoid. It uses two trapezoids and morphs them into a parallelogram.

The worksheet calls for a *straightedge.* It is simply a tool for drawing a straight line. It usually means you are not using your T-square and triangle together.

You will be finding the area of several trapezoids. Be sure you find the right w_1 and w_2. They are the parallel lines. The height h must be perpendicular to the parallel lines.

Area of Hexagons

GOALS 1. To compare areas of a hexagon calculated different ways
2. To calculate the areas of stars

MATERIALS Worksheets 39-1 and 39-2
Drawing board, T-square, 30-60 triangle, 4-in-1 ruler
Calculator

ACTIVITIES ***Area of a hexagon.*** Below is the figure of a hexagon. Write down as many ways as you can to divide the hexagon into polygons whose areas you can calculate. Share your ideas with your partner.

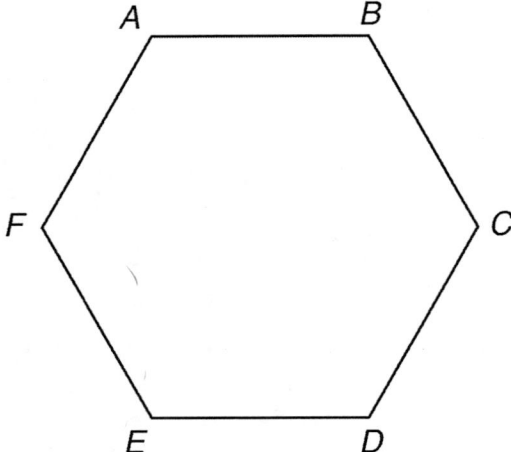

How could you find the area of this hexagon? There are at least six ways you could divide the hexagon so you could calculate the area.

Some divisions include rectangles and trapezoids.

Before you start, estimate the area in cm^2.

Worksheets. These worksheets show six ways to divide that hexagon. Use your T-square and 30-60 triangle to draw the heights of triangles. Explain your work.

In problem 6, you will need to *subtract* to find the area.

Area of stars. In one of the early lessons, you drew stars in hexagons. Now you are to calculate the areas of those stars and to find the ratio of the two areas. Do the calculations on the back of your worksheet.

A hexagonal building next to an old church in Berlin, Germany.

Lesson 40

Area of Octagons

GOALS
1. To think of ways to divide octagons in order to find the area
2. To describe the ways to find the area without actually performing the calculations

MATERIALS
Worksheet 40
Drawing board (optional), straightedge

ACTIVITIES
Finding area. When you found the area of the hexagon, you realized there was more than one solution. With octagons there are even more solutions.

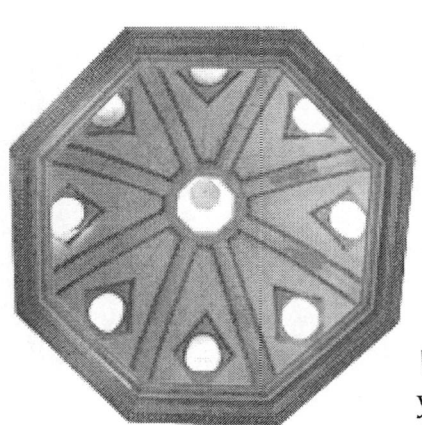

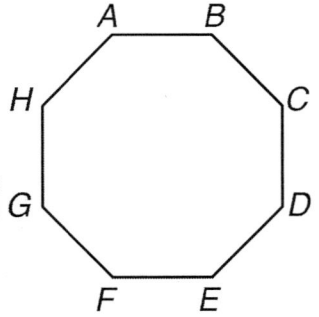

How many different ways can you find the area of this octagon?

Worksheet. You are to divide the octagon into polygons with areas you could calculate (triangles, rectangles, parallelograms, and trapezoids) in six different ways. However, you need not perform the calculations. One solution is shown below.

Ceilings of two churches in Switzerland.

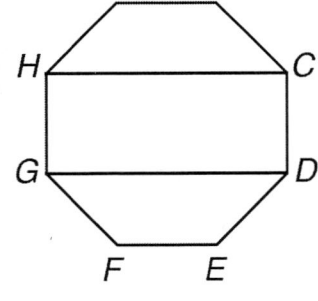

Breaking the octagon into two trapezoids and a rectangle is one solution for finding the area.

Since the area is broken into two trapezoids and one rectangle, the area is

$$A \text{ (octagon)} = 2 \times A \text{ (ABCH)} + A \text{ (HCDG)}$$

Cover of a metal container.

Octagonal window.

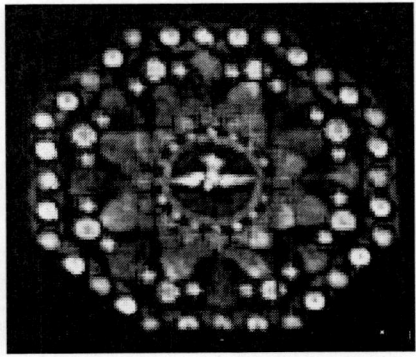

Window in St. Francis Church, San Antonio, Texas.

Lesson 41

Ratios of Areas

GOALS   1. To think about the ratios of the areas of two polygons
2. To construct a new polygon that is a given ratio of the original polygon

MATERIALS   Worksheet 41
Drawing board, 30-60 triangle, 4-in-1 ruler

ACTIVITIES   ***Some ratios of areas.*** In the figures below, rectangle A and rectangle B have the same width. That is what the slash marks mean. Rectangle A is twice the height of rectangle B. What is the ratio of the area of rectangle a to rectangle B? The answer is at the bottom of the page.

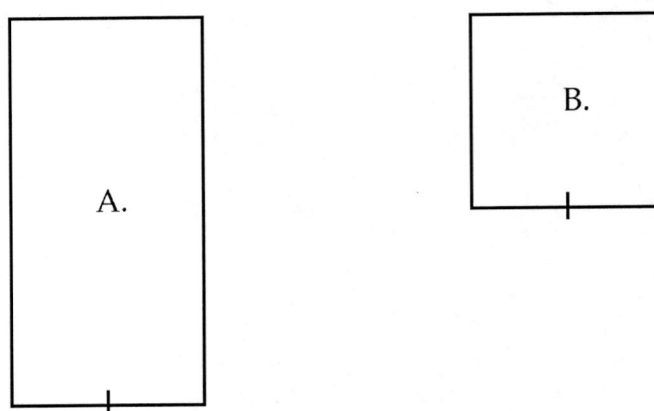

Next look at the two figures below. The polygons have the same width and the same height. What is the ratio of the area of the triangle to the area of the parallelogram? The answer is at the bottom of the page.

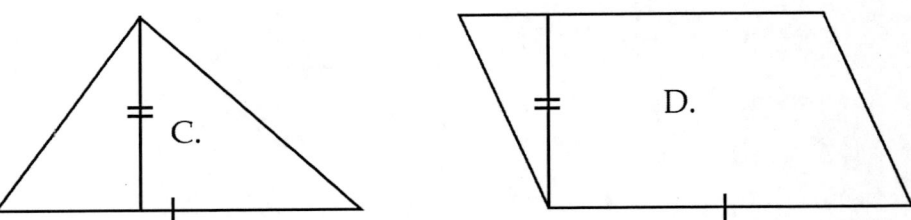

Worksheet. For this worksheet you will construct polygons that have a certain ratio of area of the given rectangle. In several cases you will need to measure the original rectangle and to measure the line segments in the new figure. [Answers: 2 to 1; 1 to 2.]

Lesson 42

Measuring Angles

GOALS
1. To learn to use a *goniometer*
2. To review the symbol for degree (°)
3. To construct figures and to measure the angles with a goniometer

MATERIALS
Worksheet 42
Drawing board, T-square, 30-60 triangle, 4-in-1 ruler
A second 30-60 triangle or 1 plain sheet of paper and scissors
Goniometer, preferably the Pro Ruler brand

ACTIVITIES
Goniometer. Shown below is a goniometer (gon-ee-OM-ah-ter). It's very unlikely you have heard of it. The word *goniometer* includes "gon" meaning *angle,* and "meter" meaning *measure,* so it's a tool for measuring angles. A goniometer is easier to use than a protractor.

> *How many angles are in a hexagon? Since "hex" means six and "gon" means angle, "hexagon" means six angles.*

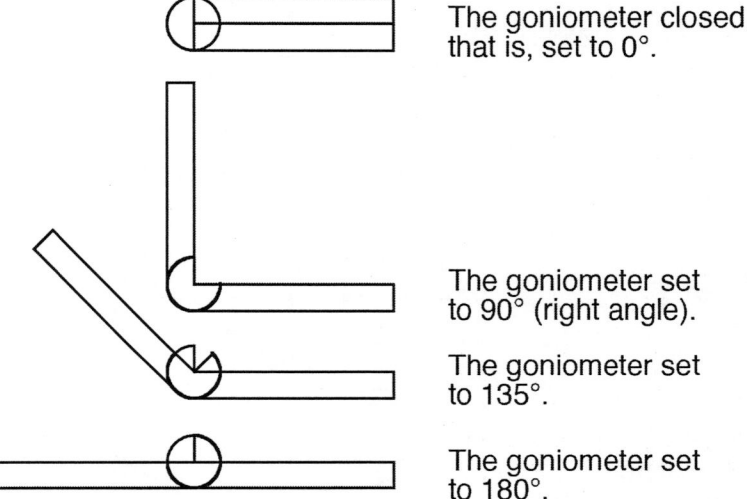

The goniometer closed, that is, set to 0°.

The goniometer set to 90° (right angle).

The goniometer set to 135°.

The goniometer set to 180°.

> *Usually, the degree symbol is skipped in writing and in equations.*

Degree symbol. The Pro Ruler brand goniometer measures angles in degrees. (Sometimes angles are measured in radians.) The symbol for degrees of an angle is the same as the degree symbol used with temperature, "°," a little circle that lines up with the top of a number.

Measuring angles. Set your goniometer to a right angle. Read 90 in the miniature magnifying glass. Try some other angles and read the results. You might even try your elbow.

Next measure the angles on your 30-60 triangle. Measuring the 60° angle is shown on the right. Are the angles what you expect?

Measuring the 60° angle on the 30-60 triangle.

> *If your goniometer should come apart, it can easily be snapped back together.*

Worksheet. For this worksheet, you need two 30-60 triangles. If you don't have a partner who will lend you one, you need to construct two 30-60 triangles out of paper. Make them the same size; they need not be as large as the plastic versions.

Use your triangles to make each requested figure. Then construct it with your drawing tools and ruler. Measure the angles in your constructed figures. The first one is done for you.

Lesson 43

Supplementary and Vertical Angles

GOALS
1. To learn the terms, *supplementary angle, vertical angle,* and *complementary angle*
2. To apply the properties of these angles

MATERIALS
Worksheet 43
Goniometer

ACTIVITIES
Supplementary angles. Below are three pairs of supplementary angles; *a* & *b*; c & *d*, and *e* & *f.* Measure the angles on the worksheet.

What is a good definition of supplementary angles? The answer is at the bottom of the page.

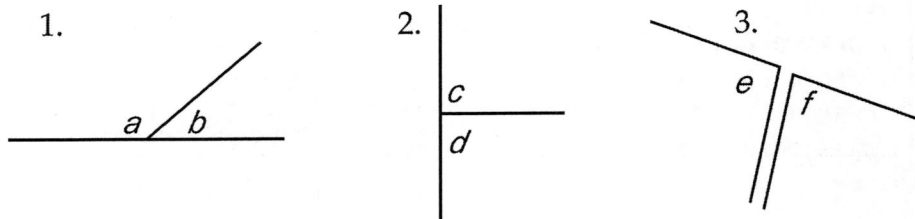

The figure shows three pairs of supplementary angles.

Vertical angles. In the left two figures below, the sets of vertical angles are shaded. You might think of them as "opposite angles."

> **Vertical angles look like "opposite angles."**

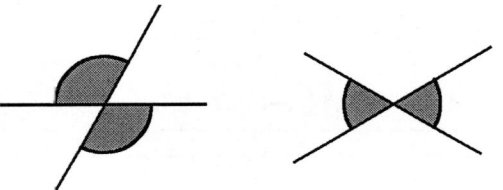 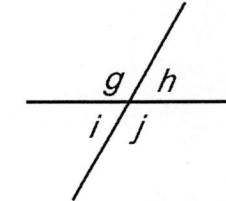

The vertical angles are shown shaded in the two left figures.

Angles g & j and h & i are vertical angles.

Do the vertical angles look congruent? Refer to the right figure above to understand why. In the figure, m∠g = 120. What is ∠h? (Remember that ∠g and ∠h are supplementary angles.) The answer is at the bottom of the page.

> **Rather than write out "measure of angle a," the abbreviation m∠a is used.**

Now find m∠j, *a*gain using supplementary angles. Also find m∠i by using supplemental angles once more. If you've thought this through carefully, you will realize that vertical angles are equal.

Complementary angles. To be complete, we must mention one other category of angles. Two angles measuring 90 are called complementary angles. *Complementary* is what is needed to make something complete. So, a complementary angle is the angle needed to make 90°. Notice that the word *complementary* has no *i*. I am not complimenting you; rather, it has two *e*'s like the word *complete*.

Worksheet. Now apply these concepts on the worksheet. [Answers: Two angles are supplementary if the measure of their angles equals 180; m∠h = 60; m∠j = 120; and m∠i = 60]

Lesson 44

Measure of the Angles in a Polygon

GOALS
1. To investigate the measure of the angles in a triangle
2. To extend that concept to the measure of the angles in a polygon

MATERIALS
Worksheet 44
Scissors and glue
Goniometer

ACTIVITIES
Worksheet. Do Exercises 1-5 on the worksheet before reading any further.

Measure of the angles in a polygon. As you discovered, one of the neat things about a triangle is that the measure of its three angles is 180.

You can quickly find the measure of the angles in a square, since it's four times 90, or 360°.

Think about the measure of the angles in a quadrilateral. If you didn't know the measure of the individual angles, you could still figure out the total sum by dividing the quadrilateral into two triangles. See the left figure below. The measure of the angles is simply 2 times 180, or 360.

Dividing a quadrilateral into the minimum number of triangles.

Dividing a quadrilateral into extra triangles, which introduces extra angles.

A tough question. You can skip this paragraph if you aren't curious or already know why you can't divide the quadrilateral into three triangles. See the figure above on the right. Actually, you can divide it into three triangles. But notice now that there are two extra angles in the middle of the quadrilateral measuring 180. So, the correct answer is 3 times 180 minus 180, or 360°, again.

Worksheet. Complete the worksheet. There is a little algebra, which will allow you to find the measure of the angles in a polygon with 20 sides.

The Weisman Art Museum at
the University of Minnesota
has many angles.

Lesson 45

Classifying Triangles by Sides and Angles

GOALS 1. To review triangle classification by congruent sides or by angles
2. To solve some angle problems

MATERIALS Worksheet 45
Drawing board, T-square, triangles
Goniometer

ACTIVITIES ***Classifying triangles.*** You probably remember the definitions of acute and obtuse angles. Here they are in mathematical symbols:

$$0° < \text{acute angle} < 90°$$
$$90° < \text{obtuse angle} < 180°$$

The charts below summarize triangle classification.

> **The slashes on the sides of a polygon mean that those line segments are congruent.**

Triangles by Sides		Triangles by Angles	
Equilateral 3 congruent	△	Acute all < 90°	△
Isosceles 2 congruent	△	Right one = 90°	◁
Scalene 0 congruent	△	Obtuse one > 90°	◿

Solving angle problems. Think about the problem below.

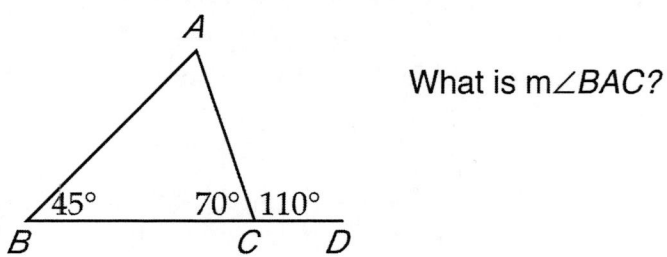

What is m∠*BAC*?

To solve the problem, keep in mind the relationships you know about angles. For example, you know the measure of the angles in a triangle is 180. But you're missing ∠ *BCA*, which you can find that by thinking of supplementary angles. The answer is at the bottom of the page.

Think of the problems as puzzles. Do a little searching and you'll think of the missing piece.

Worksheet. Exercise 10 has an important result that you will need to solve some of the problems. [Answer: 65°]

Lesson 46

External Angles of a Triangle

GOALS
1. To learn the relationship between an external angle of a triangle and the nonadjacent internal angles
2. To investigate the sum of the external angles in a triangle

MATERIALS
Worksheet 46
Scissors, glue, and tape
Goniometer, triangles

ACTIVITIES
Worksheet. Study the definitions of external and internal, and adjacent angles shown in the figure below. Then stop reading and do the worksheet before you finish reading.

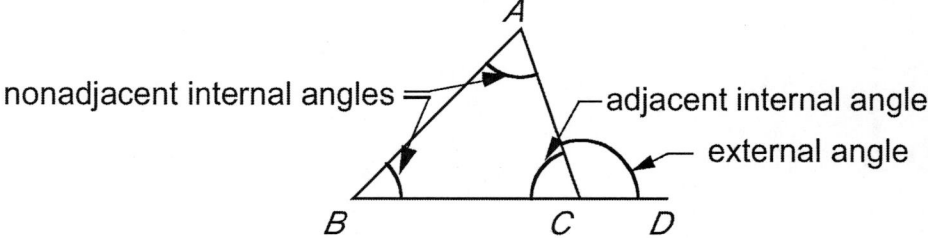

nonadjacent internal angles — adjacent internal angle
— external angle

> In math after you discover a pattern, your next task is to try to understand why.

External angle of a triangle. In question 3, you wrote the pattern you found. Simple algebra can help you understand why the external angle pattern words. Refer to the left figure below.

> You need to know. ──────▶

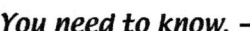
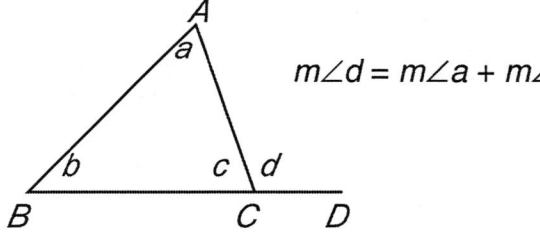

$$m\angle d = m\angle a + m\angle b$$

> The reason for the equation is given inside the parenthesis.

$m\angle c + m\angle d = 180$, (Supplementary angles)
$m\angle a + m\angle b + m\angle c = 180$, (Sum of angles in triangle)
$m\angle c + m\angle d = m\angle a + m\angle b + m\angle c$, (Equal to same number)
$m\angle d = m\angle a + m\angle b$, (Subtracting $m\angle c$ from both sides)

Consider the problem from the previous lesson. Note that it can be solved with external angles: $110 = 45 + \underline{65}$.

Measure of the sum of the external angles in a triangle. A second pattern you must have noticed is that the sum of all the external angles is 360. Algebra, again, makes it possible to understand why.

$m\angle u = m\angle s + m\angle r$, (External $\angle$)
$m\angle v = m\angle s + m\angle t$, (External $\angle$)
$m\angle w = m\angle r + m\angle t$, (External $\angle$)
$m\angle u + m\angle v + m\angle w$

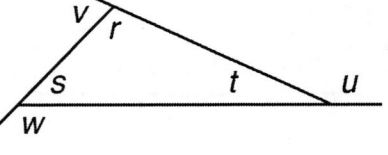

$$= m\angle s + m\angle t + m\angle s + m\angle r + m\angle r + m\angle t$$

Adding the three equations together
$$m\angle u + m\angle v + m\angle w = m\angle r + m\angle s + m\angle t + m\angle r + m\angle s + m\angle t$$
Just rearranging terms
$$m\angle u + m\angle v + m\angle w = 180 + 180, \ (m\angle r + m\angle s + m\angle t = 180)$$

Lesson 47

Angles Formed With Parallel Lines

GOALS
1. To learn *corresponding, interior, exterior, and alternate angles*
2. To learn the term *transversal*
3. To learn the relationships between these angles

MATERIALS
Worksheet 47
Goniometer (only for problems 1-2 on the worksheet)
Drawing board, T-square, 30-60 triangle, 45 triangle

ACTIVITIES
Terms. The first two lesson goals might look more like a page from an English book rather than from a math book. Once you know the terms, the math is fairly easy. It's all about a line, called a *transversal*, intersecting parallel lines. Numerous angles, eight to be exact, are formed as you can see from the figure below.

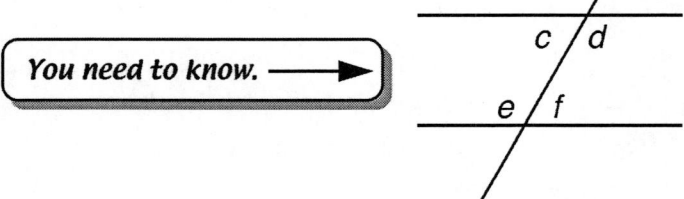

Angles formed when a transversal crosses parallel lines.

You need to know. ➡️

Two examples of corresponding angles.

You need to know. ➡️

Interior angles are inside.

Exterior angles are outside.

You need to know. ➡️

Alternate angles are on opposite sides of the transversal.

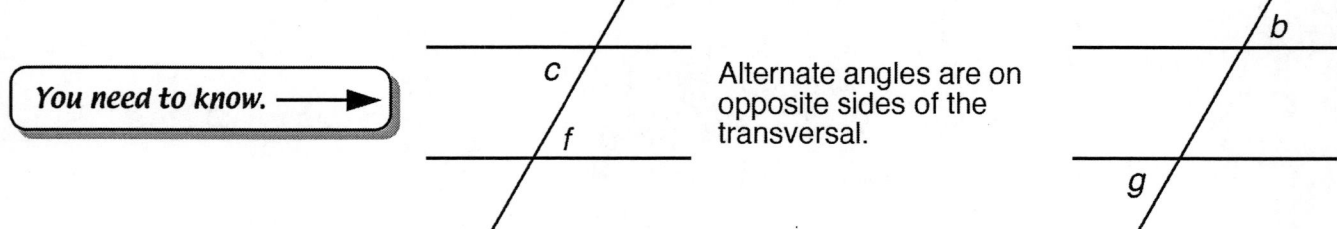

Worksheet. The worksheet asks you to find which angles are congruent and to apply the results. Use the goniometer only for the first two problems.

Lesson 48   **Triangles With Congruent Sides (SSS)**

GOALS   1. To discover the relationship between the measurements of the
               sides of a triangle
            2. To learn how many triangles can be made given three sides (SSS)

MATERIALS   Worksheet 48
            Scissors

ACTIVITIES ***Worksheet.*** On your worksheet, you need to cut out strips at the
            bottom and use them to form a triangle. See a sample triangle
            below. Consider the *inside* edges to be the sides of the triangle.

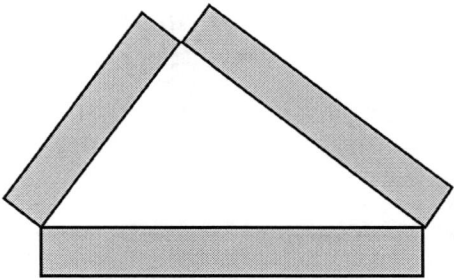

Complete the worksheet before reading further.

Sides of a triangle. Were you surprised to find that not just any
three measurements can form a triangle? We can state the results
mathematically using inequalities. If a, b, and c are the sides of a
triangle, then

$$a < b + c$$
$$b < a + c$$
$$c < b + a$$

> *Here you get a chance to use those inequalities you have been writing since first grade.*

You don't have to memorize this, just think about what happens
when to try to make a triangle.

Do you think the sides of a trapezoid could be 3, 4, 5, and 13? Does
a similar relationship apply to other polygons? The answers are at
the bottom of the page.

Congruent triangles. Two triangles are congruent when their
three sides and three angles are congruent.

You found it was impossible to make more than one triangle with
three strips. That means that two triangles that have congruent
sides are congruent. In this case you don't need to be concerned
with the angles.

This important result is often abbreviated as *SSS*, where S means
side and SSS means three sides. Therefore, SSS refers to three sides
of two triangles being equal, meaning the triangles are congruent.

You also found you could make more than one trapezoid with the
same four strips. So, other polygons having congruent sides are not
necessarily congruent. [Answers: no, yes]

Lesson 49

Other Congruent Triangles (SAS, ASA)

GOALS
1. To discover other methods for finding congruent triangles
2. To learn the mathematical meaning of the term *similar*

MATERIALS
Worksheet 49
Drawing board, T-square, triangles, goniometer

ACTIVITIES
Worksheet. Do the first four problems before reading any farther.

Congruent triangles: SAS and ASA. As you discovered with the first problem, only one triangle can be constructed with Side-Angle-Side (*SAS*) measurements given. Be sure the angle is between the sides. So, if two triangles have congruent SAS measurements, they must be congruent.

And the same thing happened with Angle-Side-Angle (*ASA*). Again, be sure the side is between the two angles. Check with your partner; all your triangles should be congruent.

Similar triangles: AAA. Something else happened in problem 3. You could construct any size you wanted. Your partner's triangle probably is not congruent with yours. In math, triangles with the same shape, but different sizes, are called *similar*. In everyday use, the word similar often means "almost alike." In geometry, similar means the same shape, although not always the same size.

*What about SSA? The two triangles below have two congruent sides, and a congruent angle that is not between the sides. Obviously, in this case the triangles are not congruent.

> **Angles with one slash are congruent. Angles with two slashes are congruent, and so on.**

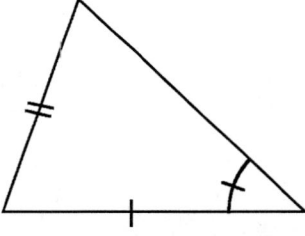
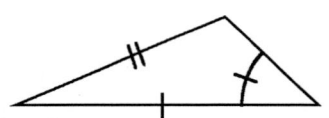

Two triangles with SSA.

Worksheet. In Problem 5, you are find whether ∆*ABC* and ∆*ACD* are congruent. Line segment *AC* is a side of both triangles so it must be congruent. In Problems 6-7 you will use some geometry you recently learned in Lesson 43.

> **You need to know. ⟶**

Summary. The sides and angles of two triangles are congruent if three of their sides and angles match one of these patterns: SSS, SAS, or ASA.

Lesson 50

Side and Angle Relationships in Triangles

GOALS 1. To learn the concept of an angle opposite a side in a triangle
2. To learn the relationship between the sides and the opposite angle in a triangle
3. To learn the terms for isosceles triangles: *vertex angle*, *base angles*, and *base*

MATERIALS Worksheet 50
Drawing board, T-square, triangles, goniometer

ACTIVITIES ***Worksheet.*** For this worksheet, you are asked to construct various triangles and measure their sides and angles. When you list the sides, write them in order from largest to smallest. Next to the lines with the measurements of the sides, list the angle.

Below is an acute scalene triangle with the measurements shown and written in order.

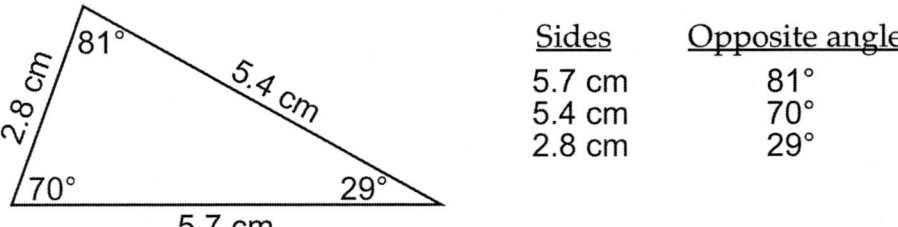

Sides	Opposite angle
5.7 cm	81°
5.4 cm	70°
2.8 cm	29°

An acute scalene triangle showing the sides and opposite angles.

Stop reading and do the worksheet. Then finish reading.

Isosceles triangles. From your worksheet, you discovered that in any triangle, the largest angle is opposite the largest side and the smallest angle is opposite the shortest side. Obviously, the side in the middle is opposite the angle in the middle.

You also discovered that angles opposite congruent sides in an isosceles triangle are congruent. This is so special that there are special names for angles and sides in an isosceles triangle.

The angle between the congruent sides is called the *vertex angle*. The congruent angles are called *base angles*, no matter how the triangle is situated. The line opposite the vertex angle, which is also the line between the base angles, is the *base*. See the figure below.

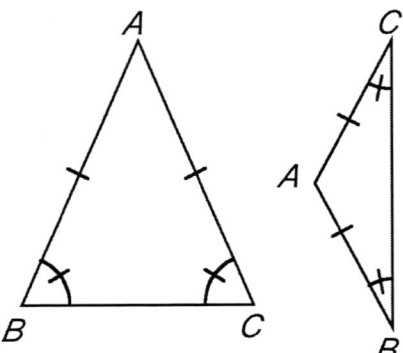

∠A is the *vertex angle*.
∠B and ∠C are *base angles*.
Line *BC* is the *base*.

Parts of isosceles triangles.

56

Lesson 51

Medians in Triangles

GOALS
1. To learn to construct the midpoint of a line (to bisect a line)
2. To learn the term *median of a triangle*
3. To construct the medians in a triangle
4. To discover where the medians intersect

MATERIALS
Worksheet 51
Drawing board, T-square, 45 triangle

ACTIVITIES
Median of a triangle. If you've ever ridden on expressway, you probably have seen the grassy strip between the two halves of the road. It usually is referred to as the *median*. You may also have learned that the median of a set of numbers in order is the middle number when they are arranged in order.

Look at the figures below and decide on a definition for a median in a triangle. The answer is at the end of the lesson.

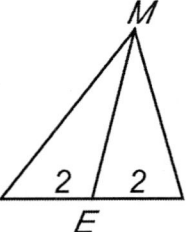 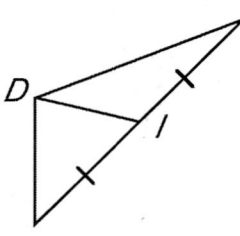 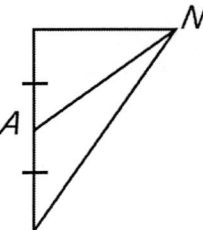

Line segments *ME*, *DI*, and *AN* are medians.

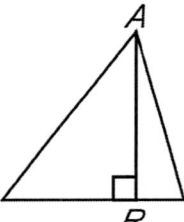 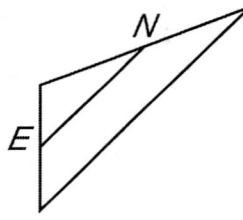 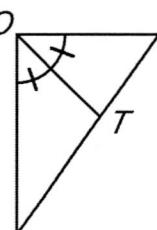

Line segments *AR*, *EN*, and *OT* are not medians.

Bisecting a line segment. Bisecting a line segment is the same thing as finding its midpoint . You have done this previously with horizontal lines. Below are the instructions for bisecting a line that is neither horizontal nor vertical.

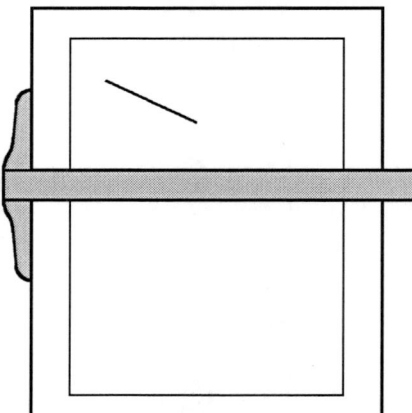 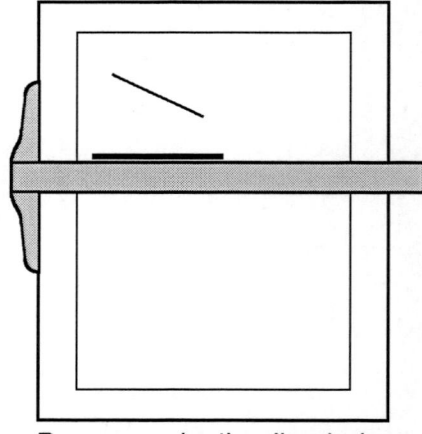

The line segment to be bisected.

Draw a projecting line below the line.

First, you project the line onto the horizontal line. It's similar to finding the line's shadow when the sun is directly overhead. Then you find the midpoint of the "shadow" and project it back to the original line. See the figures below.

You will learn another way to bisect a line in Lesson 72, using a compass.

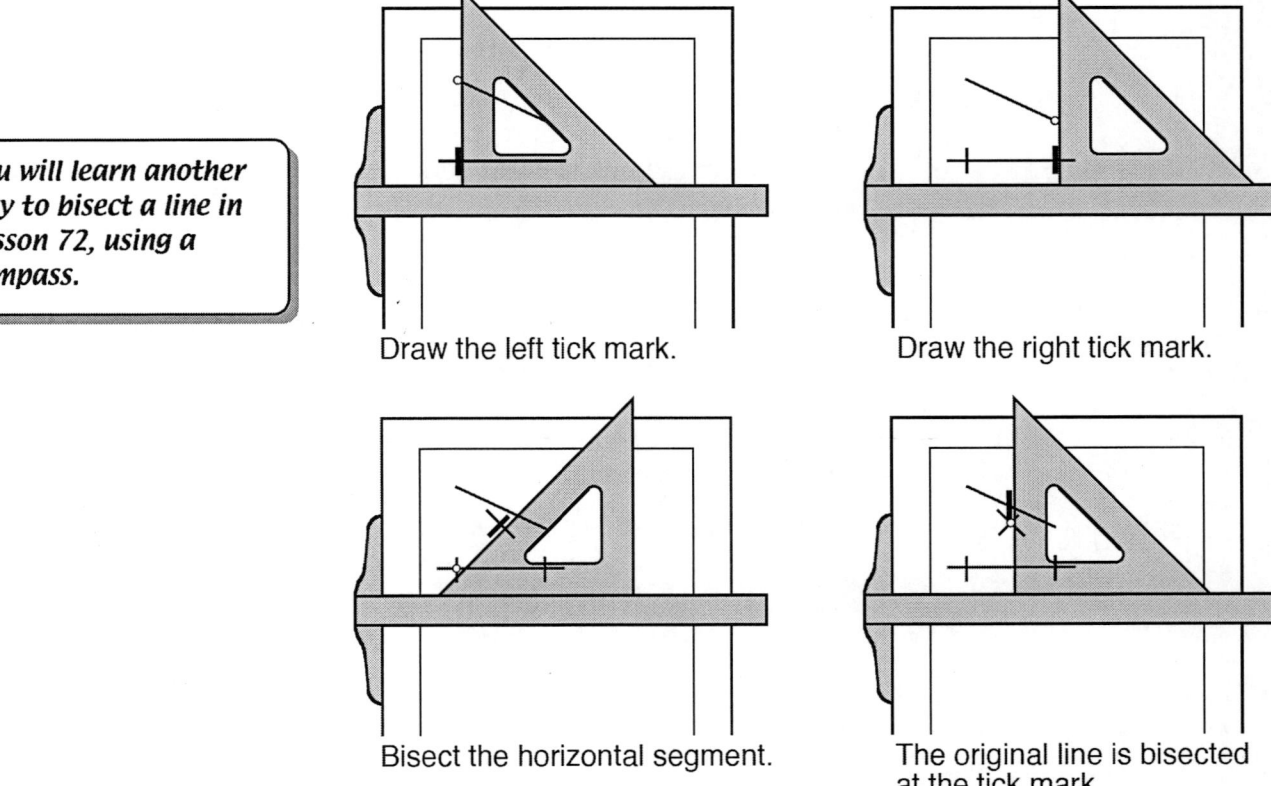

Draw the left tick mark.

Draw the right tick mark.

Bisect the horizontal segment.

The original line is bisected at the tick mark.

Bisecting a vertical line. To bisect a vertical line, skip the projecting line and proceed as you would for a horizontal line. See the figures below.

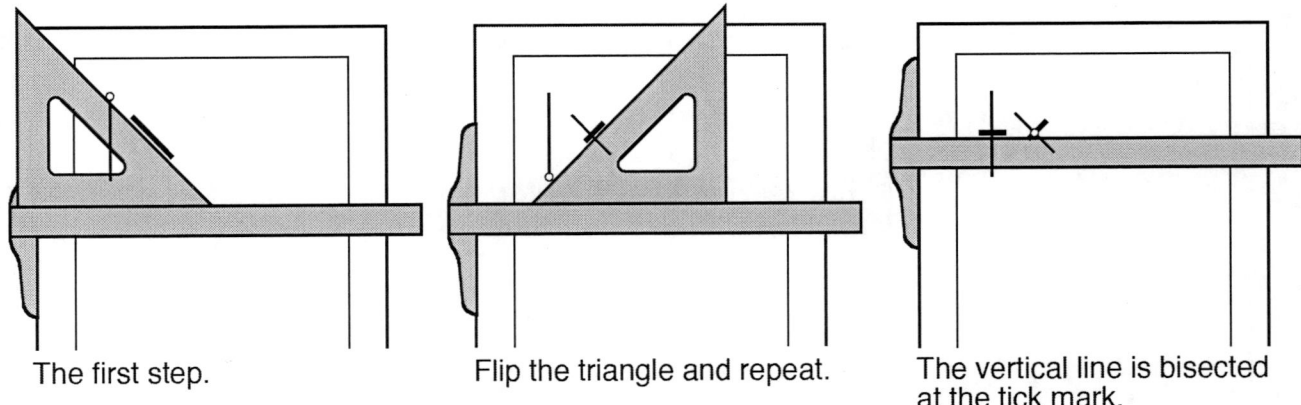

The first step.

Flip the triangle and repeat.

The vertical line is bisected at the tick mark.

Worksheet. For the triangle on the worksheet, construct the three medians. Then draw your own triangle and construct those medians. Where did your medians intersect? Check with your partner. [Answer: A median in a triangle is a line segment from a vertex to the midpoint of the opposite side.]

Lesson 52 (1 or 2 days)   **More About Medians in Triangles**

GOALS
1. To learn the term *centroid* for any triangle
2. To discover the ratio of the parts of a median divided by the centroid
3. To discover the relationships of the areas formed by the medians
4. To apply the relationships

MATERIALS
Worksheets 51 (optional), 52-1 and 52-2
Drawing board, T-square, triangles
Scissors, craft stick or long pencil with flat (not round) sides
Pencil with a new eraser
Calculator
4-in-1 ruler

ACTIVITIES
Centroid of a triangle. In the last lesson, you found that the three medians intersected at the same point, a rather unusual event. It can be proven mathematically that this always happens, no matter what size or shape of triangle you choose. The point where they intersect is called the *centroid.*

A balancing act. Cut out either of the triangles from the previous lesson (or draw a new one complete with medians and cut it out). Try to balance it on a craft stick or side of a pencil. See the left figure below. Try the other medians.

Where do you think the triangle will balance on a pencil's eraser? Try it. The answer is at the bottom of the page.

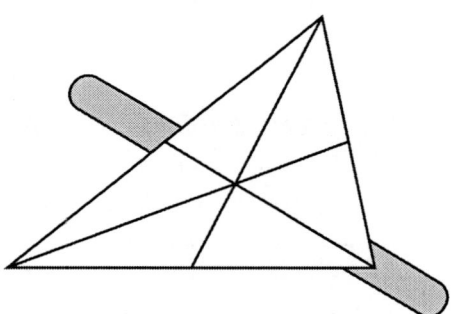

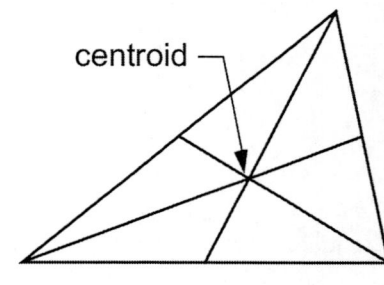

centroid

Ratios of the parts of the medians. Take a close look at a median, especially where it is divided by the centroid. It looks like each median has a long segment and a shorter segment, such as $\overline{AC}$ and $\overline{CF}$. On the worksheet, you will investigate these ratios.

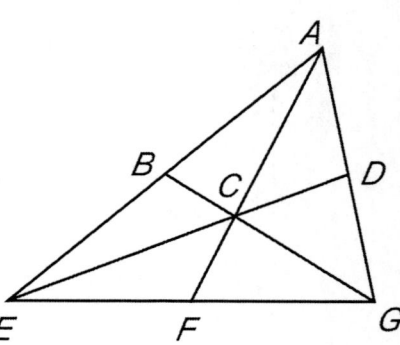

Areas of the six triangles. You will also be investigating the areas of the six triangles, such as $\triangle ABC$. and $\triangle CGF$. Before you start, which triangle do you think has the greater area?

Worksheets. Go ahead and do the worksheets. [Answer: the centroid]

Lesson 53

Connecting Midpoints in a Triangle

GOALS
1. To discover the ratio of a line segment connecting two sides of a triangle to the remaining side
2. To discover the relationship of the areas formed by the connecting midpoints
3. To apply the relationships

MATERIALS
Worksheet 53
Drawing board, T-square, triangles
Calculator
4-in-1 ruler

ACTIVITIES
Worksheet. On the worksheet, you will be connecting the midpoints in a triangle as shown below. The lines and new triangle have some interesting properties. Some lines seem to be parallel; you can check it with your drawing tools. If you've forgotten how, look back to Lesson 33.

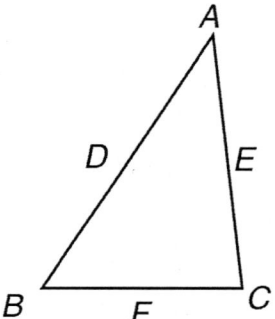 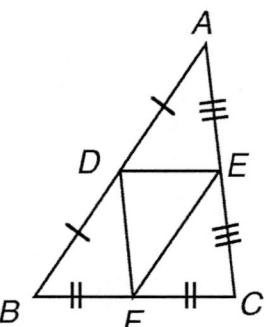

Do activities 1-6 before reading the rest of this page.

Congruent triangles. It looks like $\triangle ADE$ and $\triangle BDF$ are congruent. To see why it is, refer to the right figure above. Since line $\overline{DE} \parallel \overline{BC}$, $\angle ADE$ and $\angle ABC$ are congruent. So we've got a SAS (side-angle-side) congruence.

Actually, all four inner triangles are congruent. Use this information to answer the remaining questions.

Lesson 54

Rectangles Inscribed in a Triangle

GOALS 1. To construct a rectangle *inscribed* in a triangle at midpoints
2. To discover the ratio of the area of the rectangle to the triangle
3. To discover which inscribed rectangles have the greatest area

MATERIALS Worksheet 54
Drawing board, T-square, 45 triangle
4-in-1 ruler
Calculator
Scissors

ACTIVITIES ***Worksheet.*** For this lesson, you will be constructing rectangles. They will be *inscribed* in a triangle. A polygon is *inscribed* in another polygon if all its vertices touch the outside polygon. See the figures below. The word *scribe* means *to write.* So, to inscribe means "to write inside."

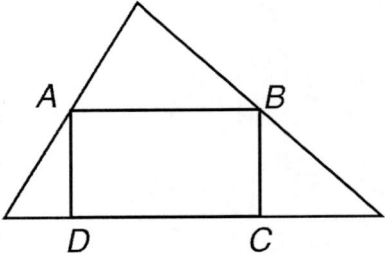

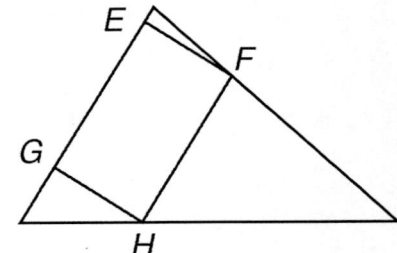

For Problem 2 on the worksheet, you will need to draw perpendicular lines in a special way. Refer to Lesson 32 if you have forgotten how to do it.

Problem 4. Compare areas with your partner. Did anyone draw a rectangle with area greater than the rectangle in Problem 1?

Complete the worksheet before reading further.

Inscribed rectangles. The last problem gives an idea about the inscribed rectangle. When two vertices of the rectangle are on the midpoints of the triangle, the area of the rectangle is half that of the triangle.

Also, the largest rectangle that can be inscribed in a triangle occurs when two vertices are at the midpoints of the triangle.

Lesson 55　　**Connecting Midpoints in a Quadrilateral**

GOALS　1. To discover the inscribed figure formed by connecting the
midpoints in a quadrilateral
2. To discover the area relationships of the figures
3. To understand the basis for this relationship
4. To learn the terms, *convex* and *concave*

MATERIALS　Worksheets 55-1 and 55-2
Drawing board, T-square, triangles
4-in-1 ruler, calculator, scissors

ACTIVITIES　***Worksheets.*** Read the comments below before doing each problem
or set of problems.

Problem 1. To do first problem on the worksheet, draw any quadrilateral. It's better if it's not fancy, such as a parallelogram. Next
find the four midpoints. Connect the adjacent points.

The figure at the right shows three connected
midpoints. When you have connected all the
midpoints in your quadrilateral, compare your
inscribed figure with your partner's figures.
Amazing!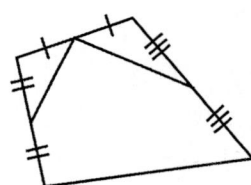

Problem 2. After you have found the ratio in Problem 2, compare
ratios with your partner. (Your areas will be different.)

Problems 3-4. This four-piece jigsaw puzzle is harder than you
think. If you need a hint, see the bottom of the page. Use this relationship to do Problem 4.

Problems 5-7. Problems 5 and 6 prepare you for problem 7. They
help you understand the area relationship in another way. First,
think about this question: You have an apple and a grape. If you eat
half the apple and half the grape, have you eaten half the fruit?

Convex and concave. The left polygon in the figure below is
convex. This means every interior angle is less than 180°. A *concave*
polygon is shown on the right. One angle is more than 180°.

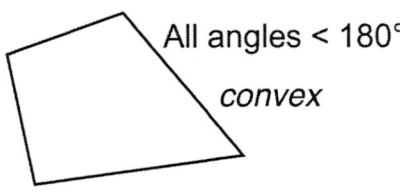

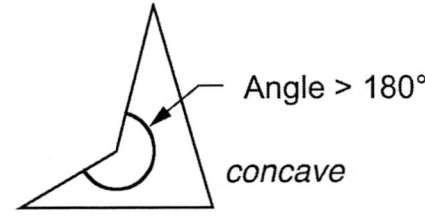

To remember the words, think that the polygons we usually work
with are *convex*, which ends like the word, *vertex*. Another way to
remember is that a *concave* polygon looks like it might have a *cave*.

Could a triangle be concave? Explain.

Problem 8. Here you are to connect the midpoints in a concave
quadrilateral. Do you think you will get the same inscribed figure?
Will the area relationship work? Try it. [Hint for puzzle: Keep all the pieces
face up. The sides of your inscribed figure are the *outsides* of the parallelogram. Fruit
answer: Yes. Note answer: No. Since all the angles in a triangle sum to 180°, no one
angle could be greater than 180°.]

Lesson 56

Introducing the Pythagorean Theorem

GOALS 1. To learn the terms, *hypotenuse* and *legs*, in a right triangle
2. To discover the relationship of the areas on the sides of a right triangle

MATERIALS Worksheet 56
Drawing board, T-square, 45 triangle

ACTIVITIES ***Sides in a right triangle.*** You will be working with right triangles for the next few lessons. The side opposite the right angle is called the *hypotenuse*. The other two sides are called *legs*. These names have been used for over four hundred years.

Usually, we name the shortest side, *a*, and the longest side, *c*. The other side is *b*. See the figure below.

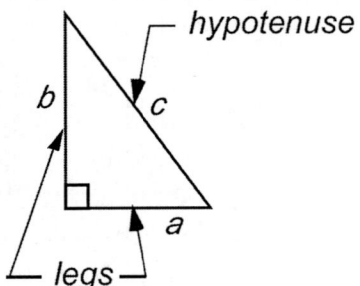

Worksheet. Do the worksheet now. Then read the next paragraph.

Pythagorean theorem. You found a special relationship between the squares on the sides of the right triangle. This is called the Pythagorean (pah-THAG-ah-ree-un) theorem. This relationship is probably the most famous and most useful result in geometry.

Lesson 57

Squares on Right Triangles

GOALS
1. To learn the term *oblique*
2. To learn to construct squares on sides of triangles
3. To investigate the sum of the squares on the sides of more right triangles

MATERIALS
Worksheet 57
Drawing board, T-square, 45 triangle
Calculator

ACTIVITIES
Pythagorean theorem. For this lesson you will be drawing squares on the three sides of a right triangle. Then you will compare their areas.

Drawing squares. Now you will draw squares on lines that are not parallel or perpendicular. Such a line is called *oblique* (oh-BLEEK). You've already learned to draw squares when the sides are horizontal or vertical. Below are the instructions.

First turn your T-square upside down. Then place the 45 triangle on the T-square. Next align them with the base of the new square. See the figure below. It's easier than it looks.

> **To refresh you memory on drawing simple squares, see Lesson 16.**

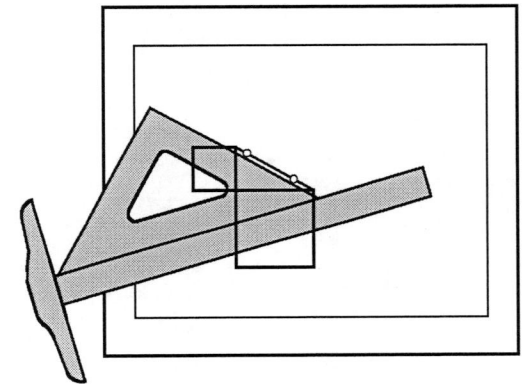

Aligning the 45 triangle and T-square to the base of the new square.

Next draw the upper edge of the square. Slide the triangle along the T-square to the correct position. Be sure to keep the T-square from moving. See the figure below.

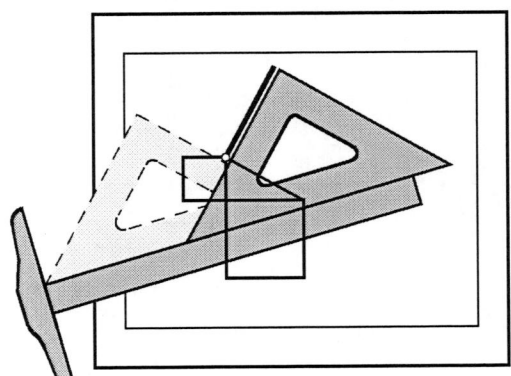

Drawing the upper side of the square.

Slide the triangle a little further and draw the lower line. See the figure on the next page. Again, don't let the T-square move.

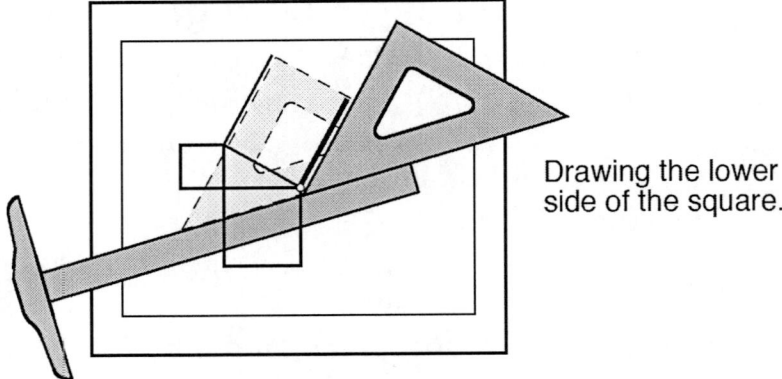

Drawing the lower
side of the square.

Now it's time to draw the tick mark, so you know the length of the side. Again, don't move the T-square. Rotate the triangle so you can pretend you're drawing the square's diagonal. Apply the tick mark as shown below.

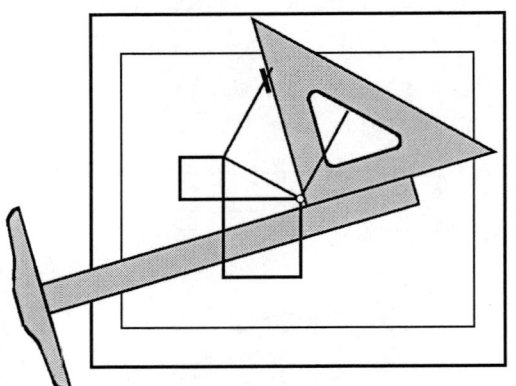

Rotating the triangle
and drawing the tick
mark.

Now you're ready for the last step. Slide the triangle so the other side of the triangle lines up with the tick mark and draw the last line. Of course, you kept the T-square from moving.

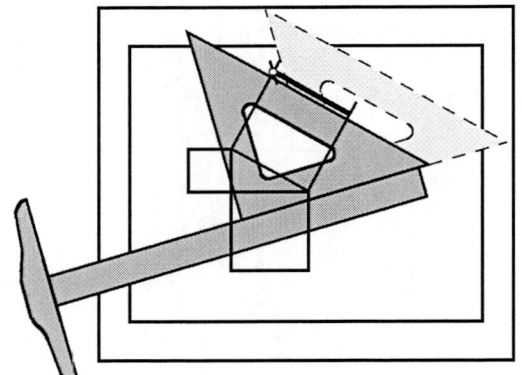

Drawing the last side
of the square.

Worksheet. When you have drawn all the squares, fill in the chart.

Lesson 58

Proofs of the Pythagorean Theorem

GOALS   1. To learn a little about its history of the *Pythagorean theorem*
2. To understand two proofs of the Pythagorean theorem

MATERIALS   Worksheets 58-1 and 58-2
Drawing board, 45 triangle,
Scissors
Calculator

ACTIVITIES   ***A short history of the Pythagorean theorem.*** Pythagoras, a Greek was one of the earliest mathematicians. He lived about 500 B.C. Pythagoras was not the first to know about the Pythagorean theorem. Four thousand years ago, Egyptians knew about the special 3, 4, 5 triangle. The Chinese knew about it at least two thousand years ago. Hindu mathematicians knew five hundred years before the Chinese.

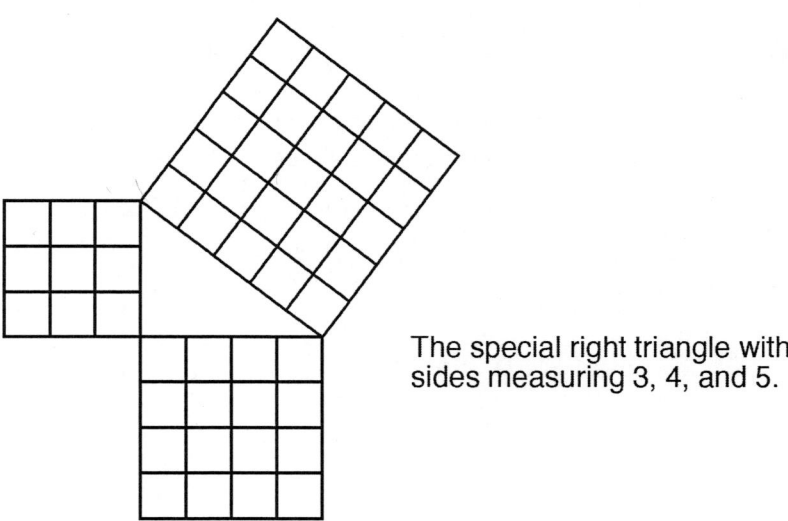

The special right triangle with sides measuring 3, 4, and 5.

> *There are over 400 ways for proving the Pythagorean Theorem. One was discovered by the 20th U.S. President, James Garfield.*

Worksheets. A proof is a set of logical reasons for learning if a statement is true. Each worksheet has a simple informal proof. These proofs only work with right triangles. The proof shows that the square on the hypotenuse equals the sum of the squares on the legs. It is usually written as

$$c^2 = a^2 + b^2$$

The figures below show the meaning of a, b, and c.

> **You need to know.** ➡️

$$c^2 = a^2 + b^2$$

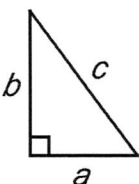

The meaning of the symbols for the Pythagorean theorem.

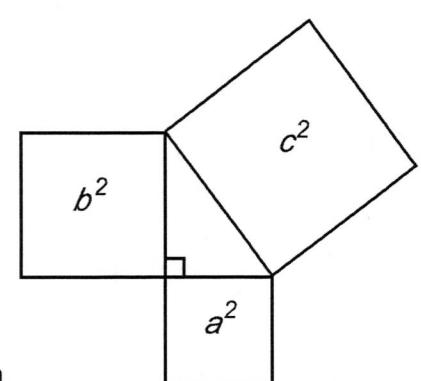

Lesson 59 — Finding Square Roots

GOALS
1. To review or learn the terms *square root, integer,* and *perfect square*
2. To review or learn the square root symbol, ($\sqrt{}$)
3. To find square roots on a basic calculator
4. To solve problems using the Pythagorean theorem

MATERIALS
Worksheets 59-1, 59-2
A basic calculator (non-scientific, not the Casio fx-300ms)
30-60 triangle, 45 triangle, 4-in-1 ruler

ACTIVITIES
Worksheets. Read the comments below before doing each set of problems on the worksheets.

Problems 1-2. These are area review problems. Instead of finding the area, you are to find the side of the rectangle. Find the lengths to the nearest tenth of a centimeter.

Square roots. When you are given the area of a square, you can find the length of a side. In a square the width equals the height, so you need to find a number that multiplied by itself gives the area. Such a number is called the *square root*. Square root is the inverse of squaring, just as subtraction is the inverse of addition.

For example, see the square below with an area of 64 cm^2. The side must be 8 cm because $8 \times 8 = 64$. The *square root* of 64 is 8.

$$A = 64\ \text{cm}^2$$

Mathematically, we write it as

$$w = \sqrt{64} = 8$$

What are the square roots of 25 and 16? What is $\sqrt{1}$ and $\sqrt{100}$? The answers are at the bottom of the page.

> **To find a square root on a calculator, enter the number. Then press the square root key, $\boxed{\sqrt{}}$.**

Problems 3-7. For problems 3-4, you already know the square roots. You will need your calculator for the other problems 5-7. The square roots for problems 3-6 are *integers,* or whole numbers. When the square root of a number is an integer, the number is a *perfect square*. Therefore, 25, 16, and 64 are perfect squares.

The answer to Problem 7 is not an integer. Round it to the nearest tenth.

Problems 8-10. To solve these problems, you will need to combine square roots and the Pythagorean theorem. Find the areas before finding the lengths of the sides. [Answers: 5, 4, 1, 10]

Solving Pythagorean theorem problems. There are so many problems that can be solved by using the Pythagorean theorem that it helps to be organized. Let's say you want to find the length of the diagonal in the rectangle below.

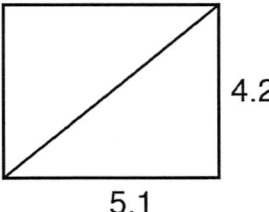

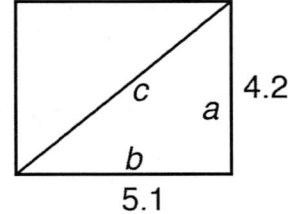

First, you may want to write in the *a*, *b*, *c*, as shown above in the second figure. Next write the formula

$$c^2 = a^2 + b^2$$

> **When using a calculator, it is not necessary to write down what's inside the parentheses.**

Then put in what you know and simplify:

$$c^2 = 4.2^2 + 5.1^2 = (17.64) + (26.01) = 43.65$$

Now find the value of *c*:

$$c = 6.6$$

Check to sure your answer makes sense.

Using your calculator. Learn to use your calculator efficiently. It will save you time and effort. Solve the above problem as follows.

Clear memory before starting. The "M" should not be visible.

> **On some calculators, you do not need to press the $=$ key.**

   Press 4.2 then $\boxed{\times}$ $\boxed{=}$. (17.64 will show.)

To place the result into memory,

   Press $\boxed{M+}$. (The "M" will show.)

To square the second number and add it to memory,

   Press 5.1 then $\boxed{\times}$ $\boxed{=}$ $\boxed{M+}$. (26.01 will show.)

To find the sum,

   Press $\boxed{MR}$. (43.65 will show.)

To find the square root,

   Press $\boxed{\sqrt{}}$. (6.6068146 will show.)

If the hypotenuse is given and you're looking for a leg, you will need to subtract. To subtract from memory, press $\boxed{M-}$.

Problems 11-20. These ten problems are in the form of a chart.

Problems 21-22. For these two problems, you are to measure the lengths of legs of your plastic triangles. Then calculate the hypotenuse. Finally, measure the hypotenuse on your triangle and see how close you are.

68

Lesson 60*Lesson 60*

More Right Angle Problems

GOALS
1. To solve more problems with the Pythagorean theorem
2. To learn the term *Pythagorean triple*
3. To solve problems by leaving the answer in square root form

MATERIALS
Worksheets 60-1, 60-2
Calculator

ACTIVITIES
Missing leg. The problem below has a missing value for a leg of the right triangle.

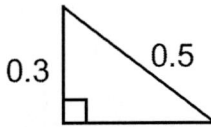

A good way to solve the problem is as follows. First, rearrange the formula so what you're looking for is on the left.
$$b^2 = c^2 - a^2$$

Then put in what you know. Be sure to *subtract* 0.3^2 from memory.
$$b^2 = 0.5^2 - 0.3^2 = 0.16$$

Finally, find the square root to find *b*:
$$b = 0.4$$

As usual, check to sure your answer makes sense.

Pythagorean triples. On Worksheet 59-2, look at Problem 11. The right triangle has sides with integers 3, 4, 5. In Problem 14 the sides are 5, 12, 13. These sets of whole numbers are called *Pythagorean triples.*

Problems 12 is a multiple of the 3, 4, 5 triangle. Each side is 2 times greater ($3 \times 2 = 6$, $4 \times 2 = 8$, $5 \times 2 = 10$). Is Triangle 13 a multiple or a new Pythagorean triple? What about Triangles 15, 16, and 18? The answers are at the bottom of the page.

Recognizing the basic Pythagorean triples can often make problem solving easier. Do you recognize the triple in the problem above?

Square root form. Sometimes the answer is best left in the square root form. That is, the solution is left with the $\sqrt{}$ symbol. For the problem below, find the hypotenuse in square root form.

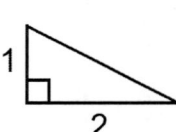
$$c^2 = a^2 + b^2$$
$$c^2 = 1^2 + 2^2 = 5$$
$$c = \sqrt{5}$$

The square root form is the exact answer. The decimal form is an approximation unless the number is a perfect square.

Worksheet. Solve the problems on the worksheet. After you've solved Problems 7-11, compare the distances and decide what makes the distance shorter. [Answers: triangle 13 is a multiple of 3, 4, 5; triangle 15 is a multiple of 5, 12, 13; triangle 16 is a new triple; triangle 18 is not a triple; 0.3, 0.4, 0.5 is a multiple of 3, 4, 5. Distance is shorter with diagonals.]

G: © Activities for Learning, Inc. 2010

Lesson 61

The Square Root Spiral

GOALS
1. To construct the square root spiral
2. To combine two spirals to make interesting patterns
3. To calculate and measure the hypotenuses in the spiral

MATERIALS
Worksheet 61-1, 61-2
Drawing board, T-square, 45 triangle
4-in-1 ruler
Calculator

ACTIVITIES
Worksheet #1. On the first worksheet you will be constructing the square root spiral. The directions are given below for the first few triangles. Make the "outside" edge of each triangle 3 cm. Continue the pattern until you have completed the spiral.

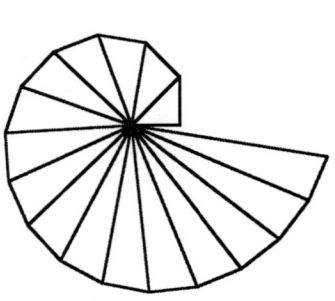

The Square Root Spiral.

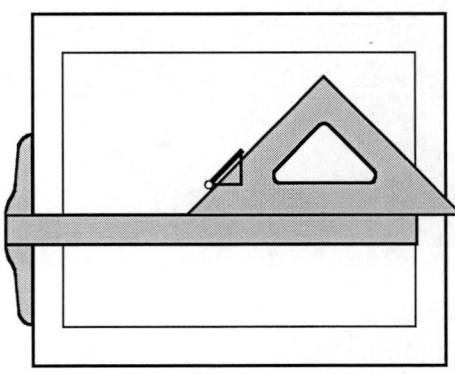

Draw the hypotenuse of the first triangle.

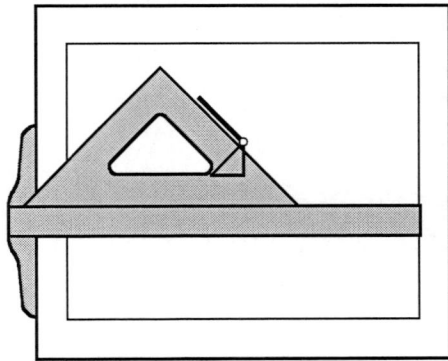

Draw leg of the second triangle.

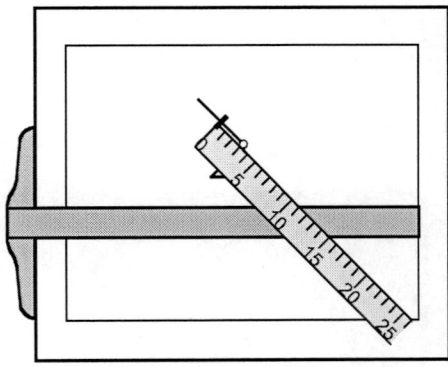

Measure 3 cm.

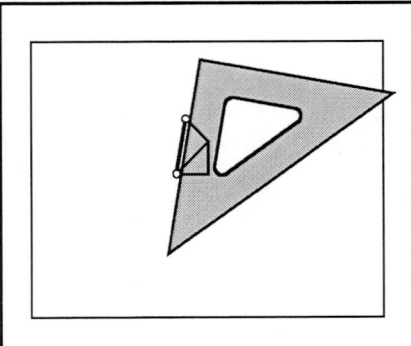

Draw the hypotenuse of the second triangle.

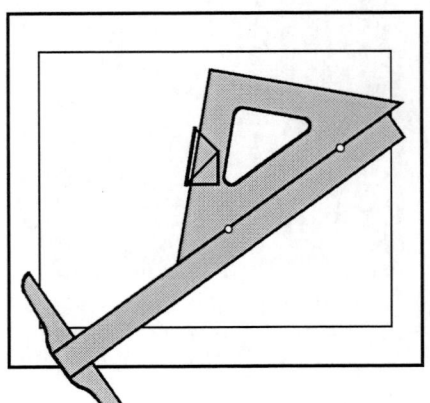

Place the T-square next to triangle.

70

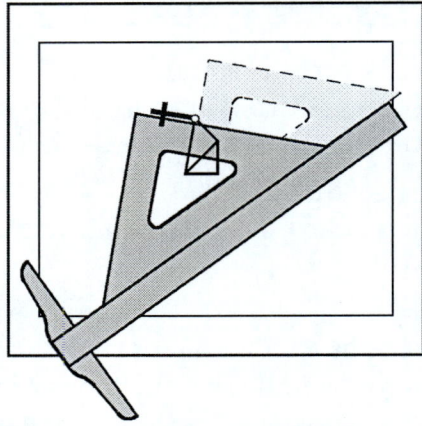
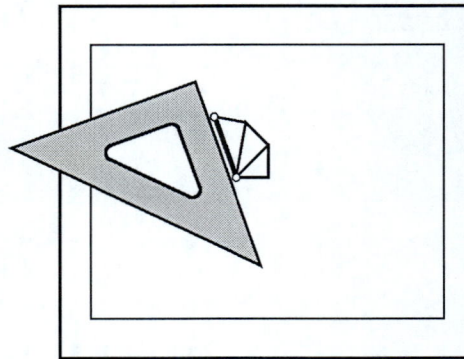

Keep the T-square still while sliding the triangle down. Draw the leg of the third triangle.

Draw the hypotenuse of the third triangle.

Continue the process until you have constructed the spiral.

Fossil from fossili-cious.com

Worksheet #2. When you complete the chart, you will see why it is called the Square Root Spiral. Solve the problems for the first column like Problem 4 on the previous worksheet.

For the second column, use your calculator to find the square roots to the nearest tenth of a centimeter.

For the third column, measure the triangles in the figure.

Combining spirals. You can see some interesting patterns by combining two spirals. Borrow your partner's spiral for a few minutes. If that is not possible, make a photocopy of your spiral. Then hold them up to a window or light. Make the designs shown below or make your own.

A snail on a tulip in Monet's Gardens in Giverny, France

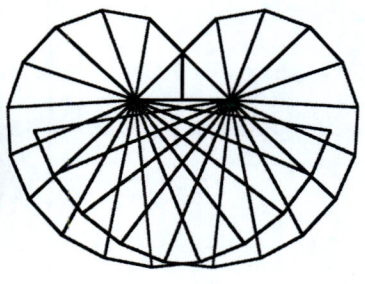
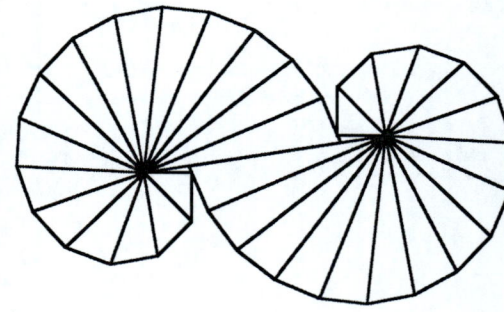
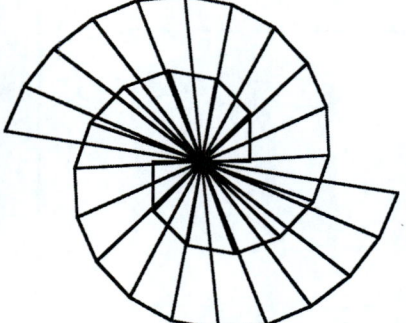
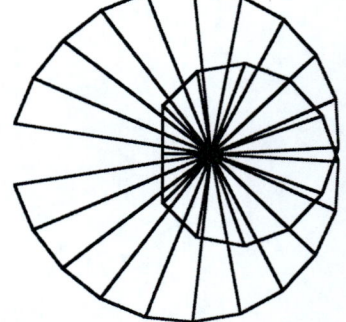

Some designs you can make by combining two square root spirals.

Lesson 62

Circle Basics

GOALS
1. To draw a circle from its definition
2. To learn the geometric meaning of *point, line,* and *plane*
3. To learn or review terms, *circumference, diameter, radius, arc, sector*

MATERIALS
Worksheet 62
4-in-1 ruler

ACTIVITIES
Worksheet. Do Activity 1 and answer Questions 2-4 on the worksheet. Then continue reading below.

Circular sculpture in Detroit, MI.

A radius is less than a diameter. The word "radius" is shorter than the word "diameter."

Circle. You drew the circle by making points the same distance from a given point. That is almost the definition of a circle. A *circle* is all the points in a plane that are the same distance from a given point.

Points. It's time to make some points about points. In mathematics, a *point* is an exact place. Although we often draw it as a dot, it is considered to have no width or height or depth. It's not something you can put your finger on.

Planes. We had to add "in a plane" to the circle's definition. A sphere is also "all the points that are the same distance from a given point." (A sphere is a shape like a ball.) Planes are pretty plain. A *plane* is a flat surface that goes on forever.

Lines and rays. A *line* can be thought of as a path made by points. A line doesn't have any thickness. As noted in lesson 1, it extends forever in both directions. A *ray*, on the other hand, is a line that starts from a point. It extends it only one direction.

Circle. Now circle back to circles. Notice that a circle is just a curved line. Mathematically speaking, it does not include the area inside. A rectangle or other polygon does not include the area either.

See the figure below for the names of the parts of a circle.

Circular design on floor.

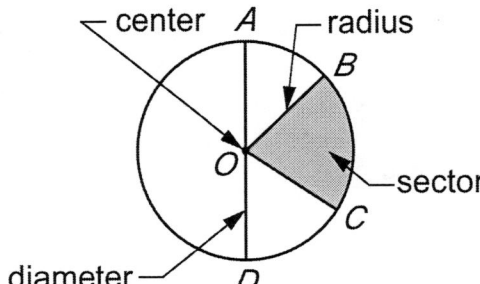

Circumference is the distance around a circle.

An *arc* is part of a circumference.

Sector is the area of a circle between two radii.

The word *radius* comes the Latin word *ray*, which describes a radius. The plural of radius is *radii* (RAY-dee-eye).

The *dia* part of *diameter* means "across." So, diameter means "measure across." This is similar to *diagonal*, which means across angles *(gon* means "angle"). Therefore a diagonal is a line between vertices.

Circular window in Lansing, Michigan

Worksheet. Complete the worksheet.

Lesson 63

Ratio of Circumference to Diameter

GOALS
1. To find the number of times a diameter will fit around a circle
2. To calculate the ratio of circumference to radius in several circles

MATERIALS
Worksheets 63-1, 63-2
At least five circular objects, such as plates, bowls, glasses, mugs, food cans or tins, vases, bicycle tire, or a circle drawn on pavement with string and chalk or drawn in sand
String
Small objects for measuring, such as beans or paper clips
Ruler

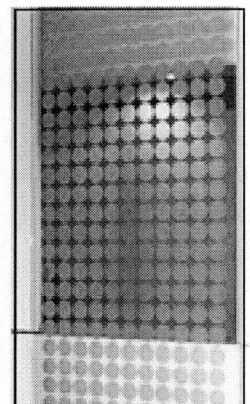

Circles on frosted glass at a York University dorm in Toronto, Canada.

ACTIVITIES

Circumference. You might have noticed that the words *circle* and *circumference* start with *circ*. So does the word *circus*. There is a connection. *Circle* originally referred to the *circus ring. Circumference* is from *circum* meaning "around," and *ferre*, meaning "to carry." All these words date back to the 14th century.

Worksheet 1. Circumference is simply the perimeter of a circle. On the first worksheet are two circles. First draw their diameters. Then measure their circumferences and diameters. Use any units you like, even beans or paper clips. Record your numbers in the chart. Then divide to find the ratio.

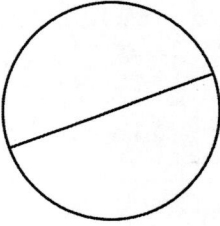

Compare your results with others who have used different units.

Other circles. Use a ruler to measure the diameter of other objects. To measure the circumference, wrap a piece of string around the object and then measure the string. Record the measurements in your chart. Find the ratio.

Circle architecture in the lobby at a York University dorm.

Worksheet 2. The second worksheet is a graph. Graph the diameters and circumferences from your chart. Also graph the diameters and circumferences from your partner and others.

Large circle (optional). Find or construct a large circle. If you can't find a ready-made one, use a long piece of string with a piece of chalk attached. Find a partner. One person holds the end of the string at the center of the circle. The other person walks around and marks the circle on the surface or in the sand.

Find out how many times the diameter fits around the circumference. You could use the length of your shoes as a measuring unit. Explain your work on the back side of the worksheet.

The London Eye in England with Big Ben, a clock tower, in the background.

A building with semi-circles in London, England.

Lesson 64

Inscribed Polygons

GOALS
1. To learn the terms, *inscribed polygon* and *regular polygon*
2. To draw polygons inscribed in a circle
3. To estimate the ratio of *C to D* in a circle by using the perimeter of an inscribed polygons

MATERIALS
Worksheet 64
Calculator
Goniometer, 4-in-1 ruler
Drawing board, 30-60 triangle, 45 triangle

ACTIVITIES
Degrees in a circle. You know that in a concave polygon (Lesson 55), one angle is more than 180. The figures below show a pencil being rotated. Each figure shows another quarter turn. In the last figure, it ends where it started. This is why a circle is said to have 360 degrees.

> Traditionally, in mathematics, angles in a circle start at the 3 o'clock position.

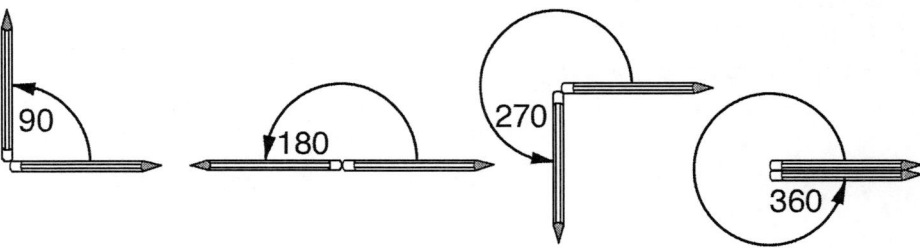

Inscribed polygon. A *polygon* is *inscribed* in a circle when all its vertices lie on the circle. A *regular polygon* is one with congruent sides and angles. See the inscribed polygons in the figure below.

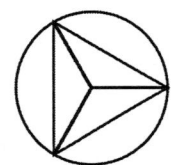   

Notice the radii in the circles. The triangle has three; the square has four; and so on. The angle between two radii in the triangle is 120. Find it by dividing 360 (the number of degrees in a circle) by 3. What is the angle between two radii for the square? For the pentagon? For the octagon? The answers are at the bottom of the page.

To complete drawing the inscribed polygon, connect the endpoints of adjacent radii.

Ratio of the perimeter to the diameter. See how the polygons with a greater number of sides come closer to the circle itself. Centuries ago people used this idea to find the circumference of a circle. This is what you will be doing in the chart on the worksheet.

Worksheet. Construct the polygons. Use your drawing board and tools for polygons with 3, 4, 6, and 12 sides. Use your goniometer and straightedge for polygons with 5 and 7 sides. Then fill in the chart. [Answers: 90, 72, 45]

Inscribed polygons on a wall.

Lesson 65

Tangents to Circles

GOALS
1. To learn the terms, *tangent* and *tangent segment*
2 To discover the relationship between a tangent and a radius
3. To discover the relationship between tangent segments
4. To solve some problems using these concepts

MATERIALS
Worksheet 65
Straightedge
Goniometer

ACTIVITIES

Tangent to a circle. In the figure on the right, the line *a* touches the circle at exactly one point. You could also say it intersects the circle at point *T* in the figure. This line is a *tangent* (TAN-jent) to the circle at point *T*.

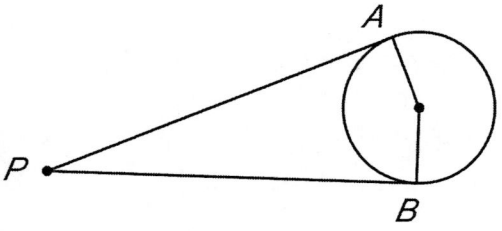

> Lines are named with lowercase letters.

Problems 1-3. You are asked to draw tangents to a circle. Then to draw a radius to the point of *tangency* (TAN-gen-see). That is the point where the tangent touches the circle. Then measure the angle the radius makes with the tangent. Compare with your partner.

Tangent to a circle. In the figure below, two tangents are drawn to a circle from point *P*. Lines $\overline{PA}$ and $\overline{PB}$ are *tangent segments*.

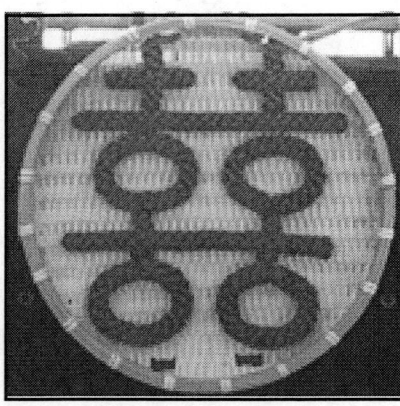

In this Chinese character, the middle two circles have a tangent line.

Problems 4-7. In these problems, you are asked to draw the two tangent segments from a point and to compare their lengths. Compare with your partner.

Problems 8-9. To solve these problems, you need to remember several things. What is the sum of the angles in a quadrilateral? What is the radius if you know the diameter? What is the angle between a tangent and radius? Do you remember some basic Pythagorean triples? [Answers: 360° (equal to two triangles); half; right angle (90°); 3, 4, 5 and 5, 12, 13 and 7, 24, 25]

Circle tangents on the building.

Lesson 66

Circumscribed Polygons

GOALS
1. To learn the term *circumscribed polygon*
2. To draw a polygon circumscribed around a circle
3. To estimate the ratio of *C* to *D* (circumference to diameter) in a circle by using the perimeter of inscribed and circumscribed polygons

MATERIALS
Worksheet 66
Calculator, goniometer
Drawing board, 30-60 triangle, 45 triangle

ACTIVITIES ***Circumscribed polygon.*** A polygon is *circumscribed* around a circle when each of its sides is tangent to the circle. The *circum-* part of the word means around. (Remember circumference?) *Scribe* means to write. So, circumscribed means to write around. See the figures below showing three polygons circumscribed around a circle.

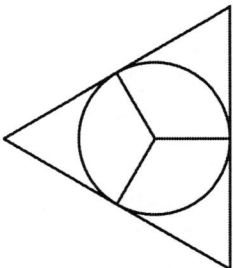 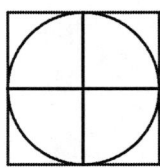 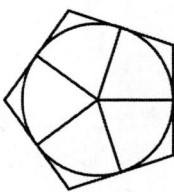

Drawing a circumscribed polygon. First draw the radii as you did for the inscribed polygons. Then draw a line perpendicular to each radius. See the figure below for a circumscribed triangle.

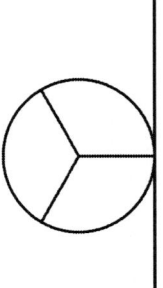 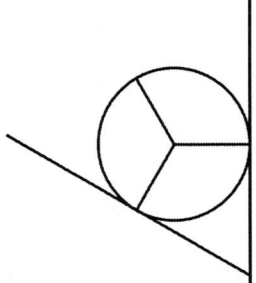 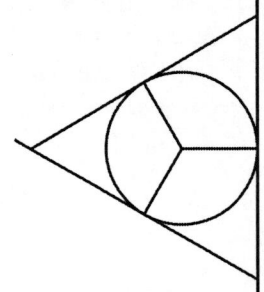

Worksheet. Construct the polygons on the worksheet. Use your drawing board and tools for polygons with 4, 6, and 8 sides. For polygons with 5 and 7 sides, use your T-square upside down and the 45 triangle. Then fill in the chart.

The final question asks you to look at the C/D ratios. Look at the ratio of the 16-sided polygon. Compare it with the ratio of the 12-sided inscribed polygon from Worksheet 64. The ratio of C/D of a circle must lie between these two numbers!

Pi, a Special Number

GOALS
1. To learn the name, *pi*, and symbol, π, for this ratio
2. To learn some simple approximations for π: 3.14, $3\frac{1}{7}$, and $\frac{22}{7}$
3. To calculate circumferences, diameters, and radii using π

MATERIALS
Worksheets 67-1, 67-2
Ruler

ACTIVITIES

Pi. In the last few lessons, you found how many times the diameter fits around a circle. The number, about 3.14, is the same for all circles. It is also the ratio of the circumference to the diameter.

Since the 1600s, this ratio has been represented by the Greek letter *p* (for perimeter), written π. We pronounce it as "pie" and spell it *pi*.

Making pi. To write the π symbol, first draw an almost vertical line. A short distance away, draw another line with a hook at the bottom. Lastly, draw the top with a hook at the left. See the figures below.

First draw a vertical line at a slight slant.

Draw a second line with a small hook.

Draw the top line, curved at the left.

> Square roots (except perfect squares) are also not exact numbers.

More pi. Pi is not an exact number. It cannot be found by dividing two numbers. If we write π = 3.1416, it is more accurate, but still not exact. If we write it as 3.14159265358979, it is closer, but still not exact. In the year 2002, pi was calculated to over 1 trillion digits!

This number has interested mathematicians for over 2500 years. Several books have been written about pi, including one for children. In a Toronto subway is a tiled wall, designed by Arlene Stamp, based on pi. There is even pi music. Lars Erickson wrote the Pi Symphony.

> *Sir Cumference and the Dragon of Pi: A Math Adventure by Cindy Neuschwander is a youth's book about pi.*

People have memorized pi to hundreds of places. According to the *Seattle Times* (2-26-95), Hiroyuki Goto, age 21, memorized over 42,000 digits of pi. It takes him over nine hours to recite it.

Values of pi. Working with more than a few digits of pi is hardly ever necessary. For most school problems, use one of these approximations: (The symbol ≈ means approximately, or about.)

$$\pi \approx 3.14, \text{ or } \frac{22}{7}, \text{ or } 3\frac{1}{7}$$

> You need to know. ⟶

Formulas for circumference. The following formula for finding circumference should make sense to you.

$$C = \pi D$$

> You need to know. ⟶

where *C* is circumference and *D* is diameter.

Since a radius is half a diameter, the formula using a radius is

$$C = 2\pi r$$

Worksheets. Now apply these concepts on the worksheets.

*Report. Research and write a report about pi.

Lesson 68

Circle Designs

GOALS
1. To review or learn the terms, *clockwise* and *counterclockwise*
2. To learn to use the mmArc Compass™
3. To make some circle designs

MATERIALS
Worksheets 68-1, 68-2
mmArc Compass™
Drawing board (The compass is easier to use if the worksheet is taped to the drawing board.)

ACTIVITIES

> Sometimes "anticlockwise" is used instead of counterclockwise.

> With his method, you can draw a circle without lifting your pencil.

> If you are left-handed, start on the right side. Draw the circle counterclockwise.

Clockwise and counterclockwise. The direction that a clock turns when you face it is called *clockwise*. The opposite direction is called *counterclockwise*. The prefix *counter-* means *contrary*.

The mmArc Compass™. With the mmArc Compass™ you can draw circles with radii from 10 to 111 mm. (But not 20 or 21 mm). If necessary, it is easy to convert the millimeters to centimeters.

Align the center of the movable part of the compass, the rotator, over the center of your circle. Next place the radius arm next to the knuckles of your non-writing hand. Draw the circle in a clockwise direction. See the left figure below.

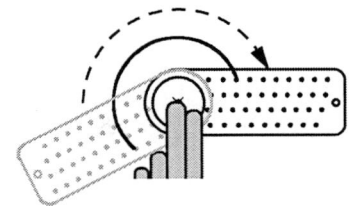 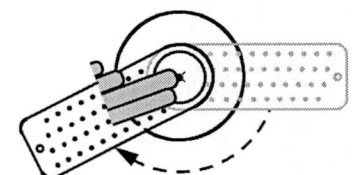

When your circle is almost complete, raise the heel of your left hand. Be sure to keep your fingers on the rotator. Continue drawing the circle until you have completed it. See the right figure above.

Design 1-2. To make the "daisy," first draw a circle. Then set the center of your compass anywhere on the circumference. Draw an arc with the *same radius*, starting and ending on the circumference. See the first figure below.

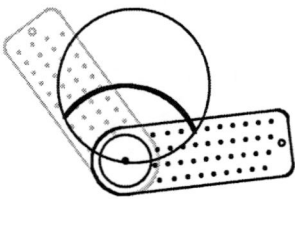 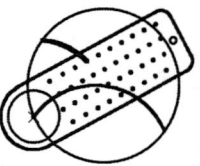  

For the second arc, position your compass at an end of the first arc. Draw another arc. See the second figure. See the third figure for the third arc. Continue drawing three more arcs. See the last figure.

Design 2 is very similar to Design 1. Instead of making arcs, draw complete circles.

Designs 3-5. Make Design 3 by drawing the arcs outside the original circle. Designs 4 and 5 are more variations.

*Original design. Draw an original design.

Circle designs adorn this window.

Circle designs on a church window.

Lesson 69

Rounding Edges With Tangents

GOALS
1. To learn to round corners, using tangents
2. To learn the terms *concentric* and *semicircle*
3. To draw a geometrical construction (a sign) from instructions

MATERIALS
Worksheets 69-1, 69-2
Drawing board, T-square, 45 triangle
mmArc Compass™
Colored pencils, red and black, optional

ACTIVITIES
Rounded corners. You, no doubt, have seen instances where two edges meet with a curve, instead of straight lines. For example, look at the corners of your mmArc Compass™ or inside your triangle. For this lesson you will construct several rounded corners.

Overview. As an example, we will round the 45-angle vertex that is inside your 45 triangle. The first figure below shows the two lines, one vertical and oblique. The next three figures show the constructions and the tangent arc. The last figure shows the arc without the constructions. It should match the inside arc of your 45 triangle.

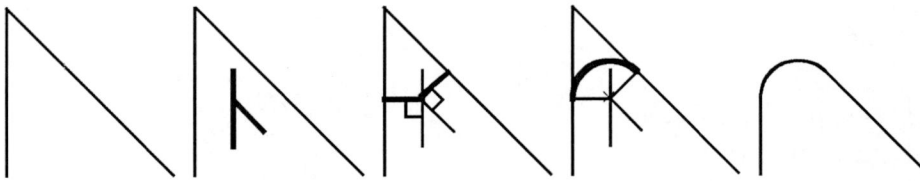

You will need to find the center of the arc. You also need the radii that are perpendicular to the point of tangency. This is necessary so you will know where to begin and end the arc. The radius of the arc is .5 cm.

Step 1. Draw lines that are 1 cm away and parallel to the sides, but inside the angle. Start by making tick marks that are 1 cm away from the sides. See the left figure below.

Next align the triangle and T-square (upside down) so the triangle is parallel to the vertical line. Then slide the triangle along the T-square to the tick mark. Draw the parallel line. See the middle figure below.

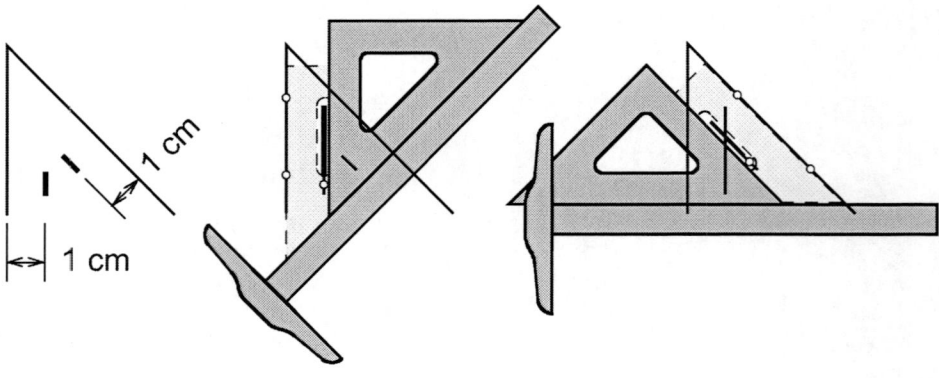

It may be helpful to draw this figure on the back of your worksheet.

Signs in Teesside, England.

Sign in Florida.

Japanese keyboard in Tokyo, Japan.

Concentric sculpture on Lake Washington in Seattle, WA.

The right figure on the previous page shows aligning and drawing the line parallel to the oblique line. Be sure your lines intersect.

Step 2. At the point of intersection, we need the lines perpendicular to the original lines–the radii. Your T-square and triangle are already in position for the oblique line. Just slide the triangle to the right. See the left figure below.

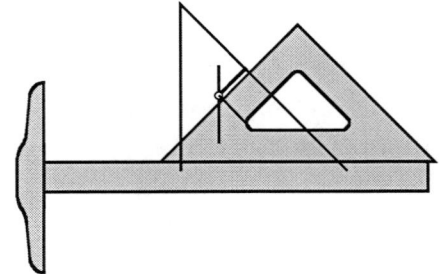
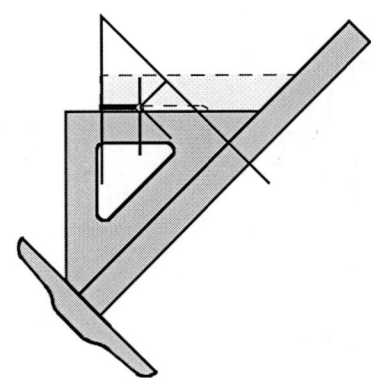

To draw the radius for the vertical line, align the tools as shown in the right figure above.

Step 3. Draw the arc using the mmArc Compass™. Look for the radius of 10 mm.

Worksheet 1. The lines on these signs are not drawn to their full lengths. That means the final step is to extend the lines to the points of tangency.

Use your T-square aligned to your drawing board for the rounded rectangles in signs 1 and 3. Use your T-square upside down for the rounded pentagons in signs 2 and 3.

Concentric and semicircle. Two circles are *concentric* when they have the same center. A roll of tape and a washer are examples. See the left figure below.

This wedding cake has three concentric circles.

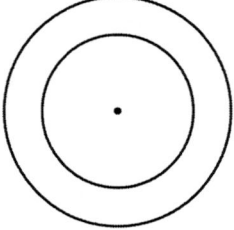
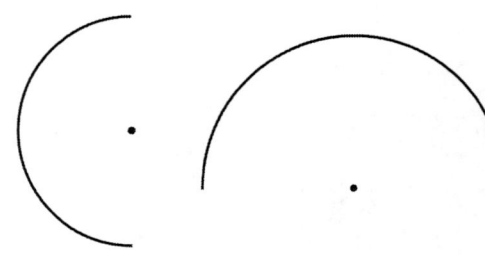

An arc that is half a circle is a *semicircle*. See the right figures above.

Worksheet 2. Here you are to construct the U-turn sign. Follow the instructions and look at the figures.

Concentric squares on a roof in Paris, France.

Lesson 70

Tangent Circles

GOALS
1. To learn the terms, *internally tangent circles* and *externally tangent circles*
2. To construct tangent circles
3. To make a circle spiral

MATERIALS
Worksheets 70-1, 70-2
mmArc Compass™
Drawing board, T-square, 30-60 triangle, 45 triangle

ACTIVITIES
Tangent circles. Tangent circles are tangent to the same line at the same point. There are two types of tangent circles. See the two examples below.

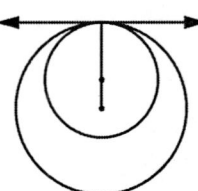

Internally tangent circles

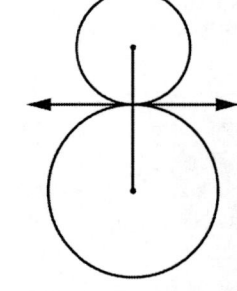

Externally tangent circles

In the left figure, one circle is inside the other. They are *internally tangent circles*. In the right figure, the tangent circles are on opposite sides of the tangent line. They are *externally tangent circles*.

Yin-yang symbol. The yin-yang is a well-known Chinese symbol. How many sets of internally tangent circles do you see? How many sets of externally tangent circles do you see? The answers are at the bottom of the page.

Yin-yang symbol

Trefoil and quatrefoil. The *trefoil* (TREE-foil) or (TREF-foil) symbol has three parts. The *quatrefoil* (CAT-er-foil) symbol has four parts. See the figures below. Both were often used in medieval architecture. Where do you see the trefoil today?

> **To make more accurate circles, keep your pencil perpendicular to the paper. Also, press against the outside of the hole while drawing.**

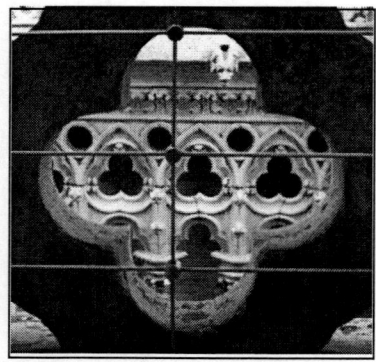

The Notre Dame Cathedral in Paris, France.

Trefoil

Quatrefoil

Circle spiral

Circle spiral. The circle spiral (above right) is made with semicircles. Starting at the center, each semicircle is greater. The spiral is smooth because the connecting circles are internally tangent.

Worksheets. Draw the figures on the worksheets. [Answers: 2; 1]

Lesson 71   **Bisecting Angles**

GOALS   1. To learn the terms *angle bisector* and *incenter*
2. To bisect an angle using a compass

MATERIALS   Worksheets 71-1, 71-2
Drawing board, 30-60 triangle, 45 triangle
mmArc Compass™
Goniometer

ACTIVITIES   ***An angle bisected.*** To bisect an angle means
to divide the angle into two equal angles. In the
figure on the right, ray AC bisects $\angle A$. That
makes $\overrightarrow{AC}$ (ray AC) the *angle bisector*.

Bisecting an angle. The figures below show
a procedure for bisecting an angle. The first
step, shown in the left figure, is to mark off equal lengths. Choose a
convenient radius and mark both rays with tick marks. If you prefer,
draw $\overarc{BD}$ instead of making tick marks.

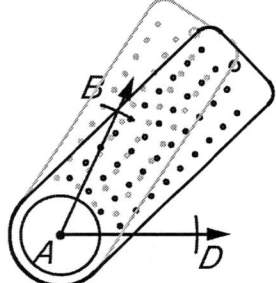
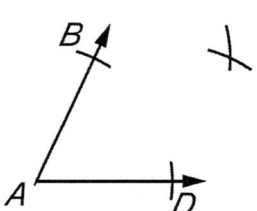
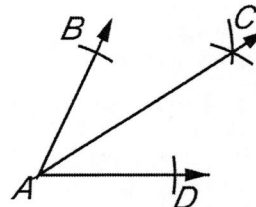

Draw equal arcs from point A.

Draw equal arcs from points B and D.

Draw $\overline{AC}$, which bisects $\angle A$.

Next draw two more arcs as tick marks as shown in the middle
figure. This time start from the tick marks at points B and D. The
radius can be different from what was used in the left figure. But it
must be the **same** for the second pair of arcs. See the figure on the
right. The angle bisector is $\overrightarrow{AC}$ drawn from point A to point C.

Why the procedure works. The key to understanding why this
works is that all radii of the same circle are congruent. Look at the
figure below. What parts of $\triangle BAC$ and $\triangle CAD$ are congruent? Which
triangle congruence proves the triangles congruent? Corresponding
parts of congruent triangles are congruent, so

$$\angle BAC \cong \angle CAD$$

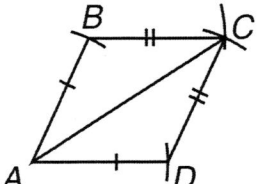

$\triangle ABC$ is congruent
to $\triangle ADC$. Why?

Worksheet. Do the first worksheet. Then continue reading.

Inscribing a circle. The point at which the angle bisectors intersect is called the *incenter* See the middle figure below. Its claim to fame is that it's the center of the inscribed triangle. See the third figure below. The name is easier to remember if you think of it as the *center* of the circle *in* the triangle.

Notice that the radii do **not** coincide with the angle bisectors. The radii must be perpendicular to triangle sides.

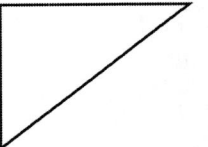 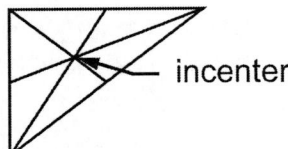 — incenter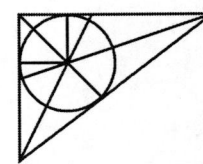

The original triangle   The triangle with angle bisectors   The inscribed circle and radii

Worksheet 2. Here you are to practice your skill at constructing angle bisectors. You will recognize the three special triangles. The radii of the inscribed circles are very special. Measure the inscribed circle before drawing it.

The circle for Problem 7 is difficult to draw with the mmArc Compass™. You will need to draw it slightly smaller.

Problem 9. The last problem doesn't fit on the page, so draw it on the back of your worksheet. Is the radius what you expected? [Answers: the three sides; SSS]

The Strasbourg Cathedral in Strasbourg, France.

Lesson 72

Perpendicular Bisectors

GOALS
1. To construct the perpendicular bisectors in a triangle
2. To learn the term *chord*
3. To find the center of a circle

MATERIALS
Worksheets 72-1, 72-2
mmArc Compass™
Goniometer, ruler
Drawing board, T-square, 30-60 triangle

ACTIVITIES
Problems 1-2. This lesson shows you another way to bisect a line segment. It is done with a compass using circle arcs. Do problems 1-2 before reading farther.

Perpendicular bisector. Since your construction divided the line segment into two equal parts, $\overline{RS}$ is a bisector. See the figure at the right. Since the bisector intersects the line segment at right angles, it is a perpendicular bisector.

Problems 3-4. The third problem asks you to find the perpendicular bisectors for all the sides of the given triangle. Problem 4 asks you to draw the circumscribed circle. Do the problems before reading further.

Circumscribed circle. The point where the three lines intersect is called the *circumcenter*. See the figure at the right.

As you discovered, the circumcenter is the center of the circumscribed circle.

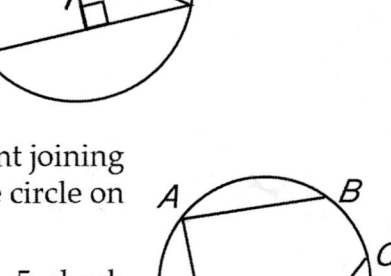
circumcenter

Chord. A chord (kord) is a line segment joining two points on a circle or a curve. In the circle on the right, $\overline{AB}$, $\overline{CD}$, and $\overline{AE}$ are chords.

Problems 5-6. After working problem 5, check with your partner and note where their perpendicular bisectors intersect.

Problem 7. Work carefully to find the center. The civil defense symbol depends upon the exact center.

The Amazing Nine-Point Circle

GOALS 1. To appreciate the nine-point circle
2. To review lines and points in a triangle

MATERIALS Worksheet 73
mmArc Compass™
Drawing tools

ACTIVITIES ***Worksheet.*** The worksheet has a scalene triangle with the amazing circle hovering over it. See the figure on the right. You will find a number of points in a triangle to be on the circle.

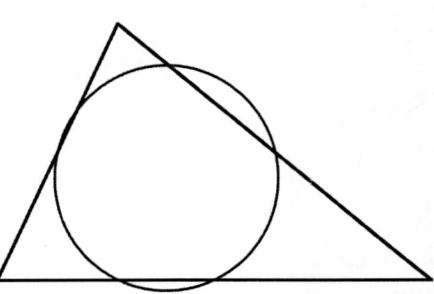

You will asked to name the points with a name like M_3. Read it as "em three." Complete the worksheet before reading any farther.

The nine-point circle. It is truly amazing that all these points are on the circumference of the same circle. The points seemed unrelated.

History of the nine-point circle. In 1820, two Frenchmen, proved that six points in a triangle were on that amazing circle. Charles Brianchon, mathematician, and Jean Victor Poncelet, an engineer and mathematician, showed that three of the points are the midpoints of the sides of the triangle. They also showed that the points where the altitudes intersect the sides of a triangle (called a *foot* or *feet*) are on the nine-point circle.

In 1822 a German mathematician, Karl Feuerbach, showed that the nine-point circle is tangent to the inscribed circle. See the left figure below. Because of his work, this circle is known in Germany as the Feuerbach circle.

A few years later, another French mathematician discovered more. Olry Terquem proved that the other three points are on the circle.

That's not all. There is another interesting feature. The radius of the nine-point circle is half the radius of the circumscribed circle around the triangle. See the right figure below.

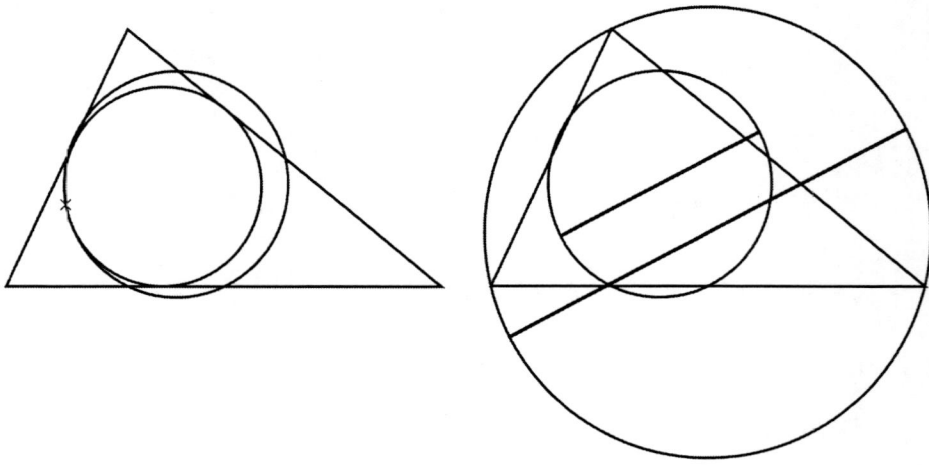

Lesson 74

Drawing Arcs

GOALS
1. To learn the term *central angle*
2. To construct figures using arcs

MATERIALS
Worksheets 74-1, 74-2, 74-3
Drawing board, 45 triangle, 30-60 triangle, mmArc Compass™

ACTIVITIES
Central angle. An arc is part of the circumference of a circle. An arc has the same measure in degrees as its central angle. A *central angle* has its vertex at the center of the circle. See the left figure below. The right figure shows just the arc.

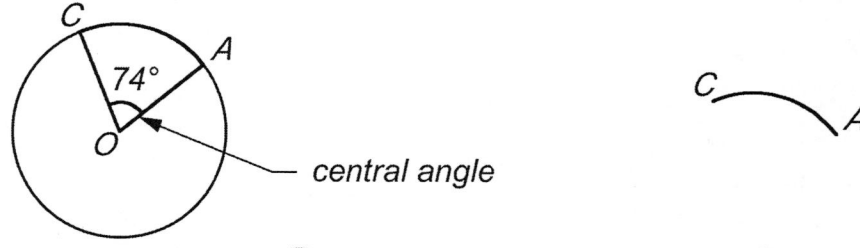

$m\angle COA = 74$ and $m\overset{\frown}{CA} = 74$. $m\overset{\frown}{CA} = 74°$.

City door in Fès, Morocco.

Worksheet 1. On the first worksheet, you will construct three versions of hearts. The first two hearts are all arcs. The third heart has straight lines, tangent to the arcs.

Worksheet 2. The figures on the second worksheet are highway signs.

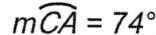

Worksheet 3. The first figure on the second worksheet is the radiation warning symbol, shown below on the left. It is composed of six 60° arcs, six line segments, and one circle. There are several ways to construct it. Plan your work to minimize erasing.

The Arc of Triumph in Paris, France

The background color of the warning symbol is yellow. The blades are usually black, although in the U.S they are often magenta.

Radiation warning symbol

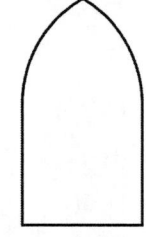
Gothic arch

The second figure, the Gothic arch, was used frequently in Gothic architecture. See the second figure on the right. You will find the arch design in doorways, arches, and windows in some old churches and college buildings.

Brooklyn Bridge in Brooklyn, New York.

Gothic barn near Hollywood, MN.

Window in Christiania, Denmark.

Lesson 75

Angles 'n Arcs

GOALS
1. To learn the terms, *inscribed angle* and *intercepted arc*
2. To discover some relationships between inscribed angles and arcs
3. To apply these relationships to solving problems

MATERIALS
Worksheets 75-1, 75-2
Drawing board, 30-60 triangle
mmArc Compass™, goniometer

ACTIVITIES
Inscribed angle. An *inscribed angle* is an angle whose vertex lies on a circle and whose sides are chords in the circle. In the circle on the right, $\angle F$ is an inscribed angle.

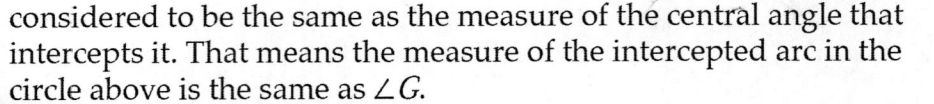

Intercepted arc. The arc between the endpoints of the chords is called the *intercepted arc.* The word *intercept* might not seem to make sense at first. But consider the derivation of the word. The first part, *inter*, means "between." The second part, *cept*, means "to take." So, the intercepted arc is the arc taken between the endpoints of the chords or endpoints of the radii.

See the circle above. $\angle G$ is the central angle. The measure of an arc is considered to be the same as the measure of the central angle that intercepts it. That means the measure of the intercepted arc in the circle above is the same as $\angle G$.

Problems 1-3. Do these problems before reading further.

Understanding problems 1-3. Did you notice that these three examples are all related? Problem 3 is a special case of an inscribed angle. Since the angle is 90°, the intercepted arc is 180°. That results in a semicircle. See the circle at the right.

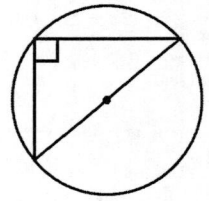

Problem 4. If you have forgotten the term you need for this problem, refer to Lesson 43. Do Problem 4 before reading any farther.

Understanding problem 4. A quadrilateral is inscribed in the left circle below. The middle circle shows the arc intercepted by $\angle U$. The right circle shows the arc intercepted by $\angle V$. Notice that the two arcs include the entire circle or 360°. So, what is $m\angle U + m\angle V$?

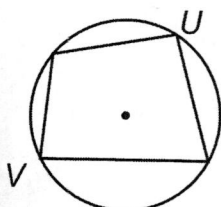 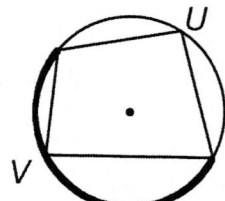 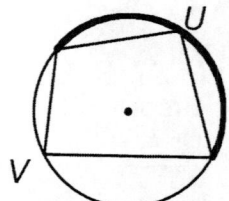

Problems 5-10. Apply these new relationships together with what you already know to solve these problems. Problem 10 shows an easier way to find the center of a circle.

Whitby Abbey in Whitby, England.

Arcs on buildings.

Lesson 76

Arc Length

GOALS 1. To learn to calculate the length of circle arcs
2. To review finding a fraction of an object
3. To review *kilometer*

MATERIALS Worksheet 76
Ruler
Goniometer

ACTIVITIES ***Arc length.*** Arc length is simply some fraction of the circumference of a circle. Consider the following problem. What is the arc length $\overset{\frown}{RST}$ in the figure below. That is, what is m$\overset{\frown}{RST}$? You know from work with fractions that since the dotted part of the arc is $\frac{1}{4}$, the solid part is $\frac{3}{4}$.

$C = 2\pi r$

$\pi = 3.1416$

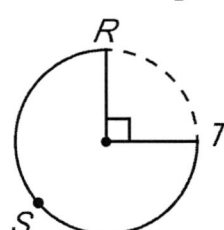

$$m\overset{\frown}{RST} = \tfrac{3}{4}C$$

Another way to think of the ratio of the arc to the whole circle is in terms of degrees in a circle. Each quarter of a circle is 90°, so three quarters is 270°. The fraction then is $\frac{270}{360}$. If you simplify the fraction, you are back to $\frac{3}{4}$.

Meters and kilometers. A meter is the basic unit in the metric system for measuring length. It is one hundred centimeters, which of course, means that a centimeter is one-hundredth of a meter. The *cent* part of the word means one-hundredth just like a cent is one-hundredth of a dollar.

A kilometer (km) is 1000 m. The prefix *kilo* means one thousand.

The word "kilometer" is pronounced two ways, either (KILL-o-ME-ter) or (kill-OM-uh-ter). The first way is used in Canada and Europe. It follows the pattern of *centimeter* and *millimeter.* The second pronunciation follows *thermometer, speedometer,* and *goniometer,* which are measuring tools.

French scientists originally designed the meter to be $\frac{1}{10,000,000}$ of the distance from the equator to the North Pole through France. Worksheet Problem 4 shows you how close the scientists were.

Practice. Solve the problem on the right before reading farther.

The $\overset{\frown}{GH}$ is twice 40, or 80°. So, the arc is the fraction $\frac{80}{360}$ of the circle. Thus, the arc length is that fraction of the circumference of the circle: $\frac{80}{360} \times 2\pi(12) \approx 16.7$ cm.

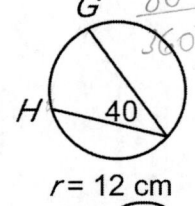
r = 12 cm
Find $\overset{\frown}{GH}$.

Worksheet. To solve the arc length problems, think about the circumference of the circle. Then find the fraction of the circle that involves the arc. It is not necessary to simplify the fraction.

*Report. Research and write a report about the metric system.

Whitby Abbey in Whitby, England.

Arcs decorate this building.

Lesson 77

Area of a Circle

GOAL 1. To understand the formula for the area of a circle

MATERIALS Worksheets 77-1, 77-2
Drawing board, 45 triangle, 30-60 triangle
Scissors
Glue or tape, optional

ACTIVITIES **Worksheet 1.** Do the worksheet before reading farther. You might want to glue or tape the pieces onto your circle.

Thinking about the area of a circle. You found that the special square with side r fits in a circle a little more than three times. The actual number of times is π, which is a little more than 3. Since the area of the square is r^2, the area of a circle is π times r^2, written πr^2 and read as "pi r squared."

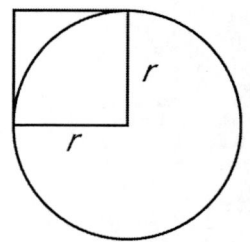

Are you surprised to find that pi is part of the area formula for a circle?

Worksheet 2. This worksheet is another way to think about finding the formula for the area of a circle. It is based on the circumference of a circle. Do the worksheet before reading further.

More thinking about the area of a circle. The second worksheet led you through a process to understand that the area of a circle is πr times r, or πr^2.

There is another way to think about the width of the parallelogram. See the figure below. The width is half the circumference of the circle because it is formed with half the sectors. You know the circumference of a circle is $2\pi r$. So half the circumference is πr. The height of each sector is r. Since the area of a parallelogram is wh, the area of a circle is πr^2.

A circular window on a church.

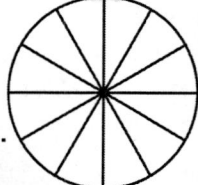 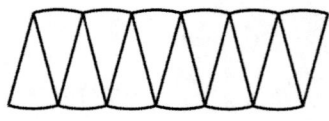

In the figure below is the same circle, divided into 24 sectors. See how much flatter the arcs appear.

This church has beautiful circular windows.

 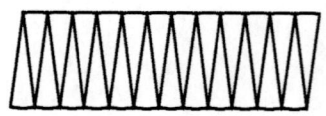

Lesson 78

Finding the Area of a Circle

GOALS
1. To learn a method for finding the area of a circle
2. To practice finding areas

MATERIALS
Worksheets 78-1, 78-2
Ruler
Calculator
Scientific calculator, if available

ACTIVITIES
Finding the area of a circle using the special square. Using the special square makes finding areas of circles easier for many students. See the figure on the right. It also helps prevent confusion between radii and diameters.

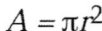

For example, what is the area of the circle on the right to the nearest hundredth of a centimeter? First, find the area of the special square. It is 1 cm². Next, there are π (a little more than 3) of these squares that will fit in the circle. Use 3.14 for π. So, the area is π × 1, which is 3.14 cm².

Using the formula for the area of a circle. Finding the area of the circle above using the formula is as follows:

$$A = \pi r^2$$
$$A = 3.14 \times 1^2$$
$$A = 3.14 \text{ cm}^2$$

Old gears.

Problems 1-3. These problems show what happens to the area when the radius is doubled. You will see also review what happens to the circumference after the doubling.

Problems 4-5. People who fly over Nebraska and surrounding states or over southwest Canada will see circles on farm fields. These fields are irrigated, or watered, by a pipe on wheels that moves like a hand on a clock.

Problem 6. Using π for pi makes it easier to see the relationship between the two areas for this problem.

Problem 7. This problem shows different answers for the area of a circle depending upon the value used for π.

The first value for pi in the chart is accurate to millionths place. You will find it on a scientific calculator.

The second value was developed by Tsu Ch'ung-chih, a Chinese worker in mechanics, about the year 470 A.D. You can remember it by thinking of the pattern,

113355

and then dividing the last three digits by the first three: $\frac{355}{113}$

The third value, 3.14, is the common everyday approximation.

The fraction, $\frac{22}{7}$, for π was very common before calculators. It is seldom used today except on tests. It makes calculations easy if the radius is a multiple of 7. Find *C* or *A* for a circle with radius = 7.

Octopus in a Tokyo fish market, in Tokyo, Japan.

Lesson 79

Finding More Areas

GOALS 1. To calculate areas of figures involving circles
2. To work with doubling perimeter, circumference, and area

MATERIALS Worksheets 79-1, 79-2
Ruler
Drawing board, 45 triangle, 30-60 triangle
Calculator

ACTIVITIES **Problem 1.** This problem has an interesting solution.

Problem 2. Find the area of the inside figure is not hard. Just look for squares and ways to make squares.

Problem 3. Window companies need to know how much area is glass in order to calculate heating or cooling losses.

Problem 4. Break up this figure into rectangles and sectors of circles. Then find and add the individual areas.

Finding a radius. If you know the area of a circle, think how you could find the radius. For example, the area is 30. Find the radius.

$$A = 30 = \pi r^2$$

To find r^2, divide by π.

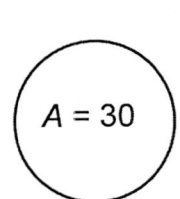

$A = 30$

$$r^2 = \frac{30}{\pi} \approx 9.55$$

To find r, take the square root.

$$r \approx 3.09$$

Worksheet 2. First measure the square or circle. Then calculate the side or radius you need to draw the new figure. Finally, draw the figure.

Problem 6. You could construct the new square and then measure the side. Or, try it both ways and compare your results. Does the new square look like it has double the area of the original square?

Problems 7-8. Does the new circle in Problem 7 seem to have double the circumference of the original circle? Does the new circle in Problem 8 seem to have double the area of the original circle?

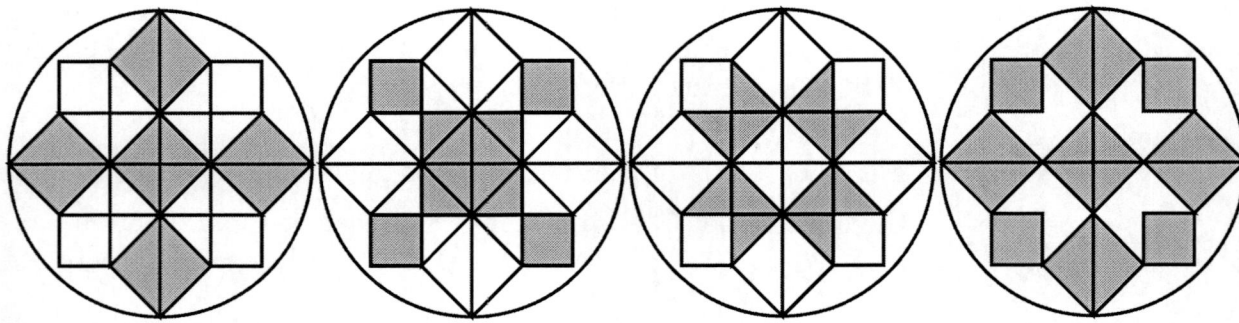

Same design with different shading.

Lesson 80

Pizza Problems

GOALS
1. To learn the meaning of *per* and *unit cost*
2. To work with tenths of a cent
3. To solve problems comparing prices

MATERIALS
Worksheet 80
Calculator

ACTIVITIES

Which is cheaper? How would you solve this problem: You want to buy six identical gifts. Store A sells them at three for $29.50 and Store B sells them at two for $18.75. Which store has the better price? Solve it before reading any farther. The answer is at the bottom of the page.

One way to solve the problem is to find out the cost for a single item at each store. This is called the *unit cost.* To find the unit cost at Store A, you divide $29.50 by 3 and to find the unit cost at Store B, you divide $18.75 by 2.

Worksheet. You can solve these problems using only a calculator without writing anything down (except the answer).

Problem 1. When you find these areas, remember that pizza dimensions refer to diameters. The symbol (") means inches.

Using "per." The phrase "Price/in^2" is read as "price *per* square inch." *Per* means "for each" or "in each." For example, "I study math five hours per week," or "My car gets 40 miles per gallon (U.S.)," or "My car's fuel consumption is 58 liters per 100 km (Canada)."

A tenth of a cent. Even though we cannot pay in tenths of a cent, calculating to the tenth of a cent gives a more accurate results. For example, suppose you want to buy some pencils costing 2 for 25¢. If you buy one, you round up the 12.5¢ and pay 13¢. If you buy 200 pencils at 2 for 25¢, you pay $25.00. If you buy 200 at 13¢, you pay $26.00. A half-cent can make a difference.

Problem 2. To find the Price/in^2, remember that it means price per area, so you need to divide. Montreal prices are in Canadian money and the Florida prices are in U.S. money.

Problems 5-6. First think about what the price of the new size pizza should be. Then check which price listed is closest.

Problems 7-8. Here you are to consider buying several pizzas. The areas will be not exact, but approximately the same.

> **Be aware that six and a half cents is written as 6.5¢ or $0.065. If you write .065¢, it is less than a penny.**

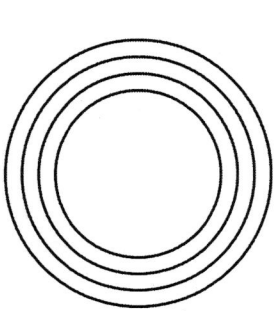

10"   12"   14"   16"

Lesson 81

Revisiting Tangrams

GOALS
1. To construct a *tangram*
2. To work with tangram shapes and draw the results
3. To review some basic concepts

MATERIALS
Worksheets 81-1, 81-2
Drawing board, T-square, 45 triangle
A set of tangrams

ACTIVITIES
Tangram. The history of the tangram is not known. It is an ancient Chinese puzzle consisting of seven pieces, which form a square. See the figure at the right. The pieces can be rearranged into hundreds of different shapes.

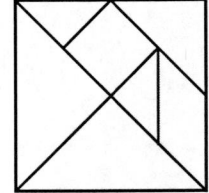

Worksheet 1. The first worksheet asks you to construct a tangram using your drawing tools. The only difficult part is bisecting the top line (finding the center). If you truly need help, look at the end of this lesson for a hint.

The remaining questions point out some interesting features of the pieces and review some basic geometry.

Worksheet 2. The worksheet has outlines of a collection of tangram pieces. You are to find tangram pieces that will make that shape. Then construct your arrangement. No two solutions can be congruent. Be sure you can justify every line you draw. No guessing.

In problems 1-4, the number of pieces to make the shape is given. For problems 5-12, you will always need four pieces.

Two examples are given below. The outlines are on the left and the solutions are on the right.

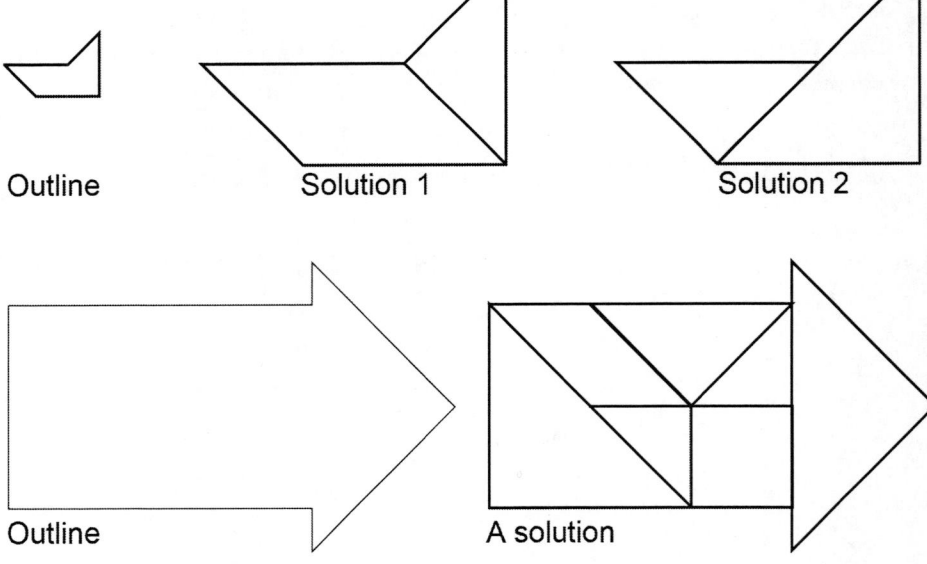

Outline   Solution 1   Solution 2

Outline   A solution

Paper folding. Make a tangram by folding a piece of paper. First fold it into a square. Then make the other folds and cut it out. [Worksheet 1 hint: It is directly above the intersection of the square's diagonals.]

Lesson 82

Aligning Objects

GOALS
1. To align objects along a horizontal or vertical line
2. To do the inverse
3. To practice visualizing

MATERIALS
Worksheets 82-1, 82-2, 82-3
Drawing board, T-square, 45 triangle
2 sets of tangrams
Colored pencils or crayons, optional

ACTIVITIES
"Align," a computer command. Many computer programs have a series of commands for *aligning,* or lining up, objects. They are called Align Left, Align Right, Align Top, and Align Bottom or something similar.

Look at the square and parallelogram tangram pieces, shown in the center of the figure below. If you slide them straight up to the *Align Top* line, they will look like the figure at the top line. If you slide them left, they will look like the figure at the Align Left line. Similarly, you can align them to the right or bottom.

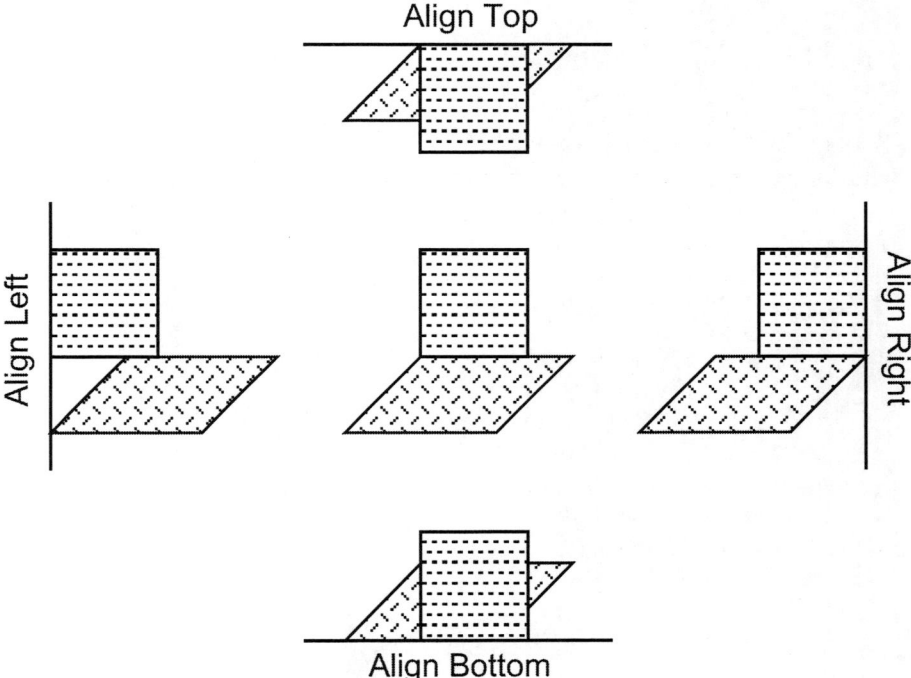

These figures show the square stacked on top of the parallelogram. If you stack the parallelogram on top of the square, the figures will look slightly different.

Worksheets 1-2. Worksheet 1 has three separate objects for you to align to the four lines. Worksheet 2 has four tangram pieces that will overlap when you align them. You decide the order to stack them. To keep track, you might want to color each shape a different color.

Worksheet 3. This worksheet has the objects aligned to the right and bottom. Your job is to visualize the tangram pieces and draw them before they were aligned.

The upside down V's on this window have a bottom alignment.

94

Reflecting

GOALS
1. To understand *reflection, image,* and *line of reflection*
2. To construct images about lines of reflection at various angles
3. To flip objects similar to the computer commands, *Flip Horizontal* and *Flip Vertical*

MATERIALS
Worksheets 83-1, 83-2
Geometry reflector, optional
Drawing board, T-square, 45 triangle
Two sets of tangrams

ACTIVITIES
Line of reflection. When you look into a mirror, you see a reflected *image* of yourself. You know your image is not the way you really look. Your right eye is on the left and your left eye is on the right.

On paper an object may be reflected across a *line of reflection*. This line of reflection, sometimes called the *mirror line*, will be shown here by a centerline. To indicate a centerline, designers use a special dashed line, a long dash alternating with a short dash. See three in the figures below. Also, see the object (two triangles) reflected three different ways.

An example of reflection found in nature.

This floor design has many lines of reflection.

A reflection along a vertical mirror line at the Durham Cathedral in Durham, England.

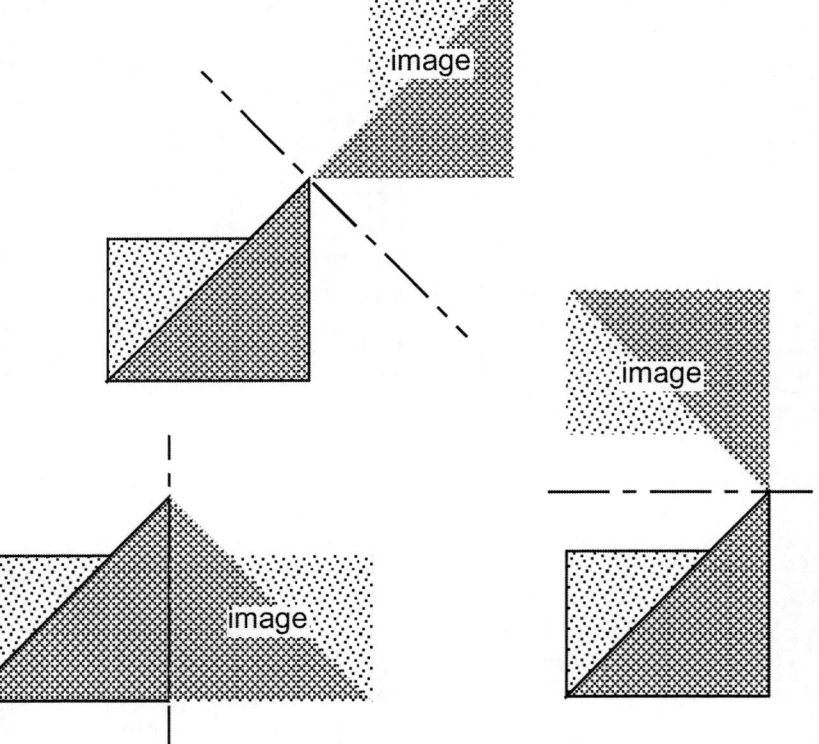

Geometric reflector (optional). A geometric reflector is reflective and also transparent.

Place the geometry reflector along the lines of reflection in the figure above and see both the reflection on paper and the reflection in the geometry reflector.

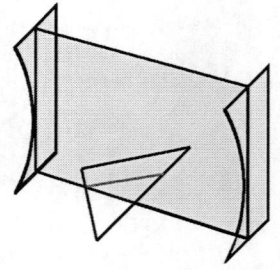
A geometry reflector.

Worksheet 1. For this worksheet, you will be reflecting the same object across nine different lines. Study the figures shown below to learn some general techniques for constructing the reflections. You may find it helpful to use your tangram pieces to construct the original figure and the reflection. To check your work, use a geometry reflector, if available.

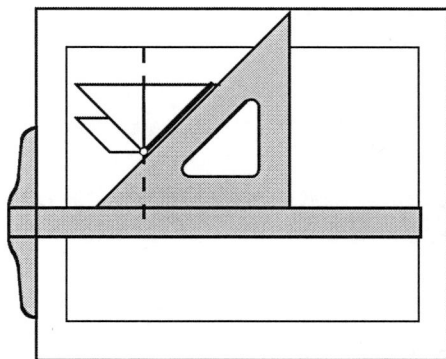 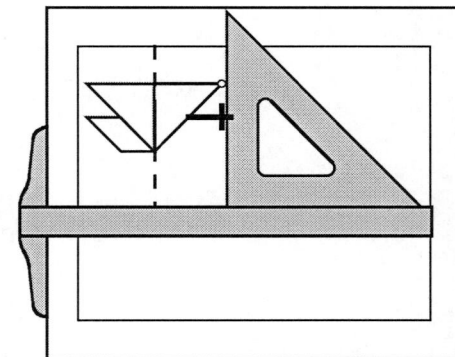

Flip, a computer command. Some software has *Flip Horizontal* and *Flip Vertical* commands. Flip horizontal means to flip about a vertical line. Flip vertical means to flip about a horizontal line.

Worksheet 2. On this worksheet, you will be flipping the same object both horizontally and vertically.

*Challenge. Take three or four tangram pieces and arrange them with the edges touching, such as the examples on the worksheet. Without any measuring, construct your design and its reflection.

A plant on the Big Island in Hawaii.

These bricks reflect along a vertical mirror line.

Lesson 84

Rotating

GOALS
1. To learn the mathematical meaning of rotation
2. To construct rotations at various angles

MATERIALS
Worksheet 84
Goniometer
A set of tangrams
Drawing board, T-square, 45 triangle

ACTIVITIES
Rotating. A clock is a good example of rotation. Both the hour and minute hands rotate about the center of the clock. The hands move in a clockwise direction. However, when we discuss rotations mathematically, we start with a horizontal ray extending right and measure the amount of rotation counterclockwise. So, for a clock to behave mathematically, the hand would start at the 3 o'clock position and travel backward.

Rotating the ship. Build the ship shown below in the left figure with four tangram triangles and tape them together.

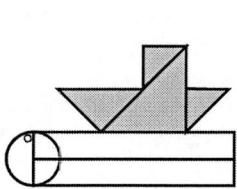 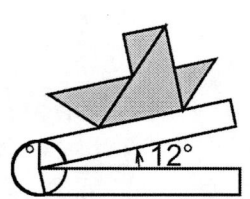 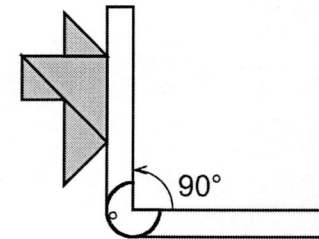

Then tape the ship to the upper arm of the goniometer. Hold the lower arm of the goniometer still with your right hand. Use your left hand to rotate and upper arm of the goniometer with the attached ship. See the middle figure above.

Keep rotating to 90° as shown in the right figure above. (The seas are getting very rough.) Continue rotating to 180°. (Disaster.) See the left figure below.

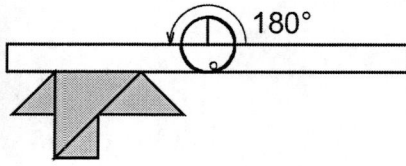 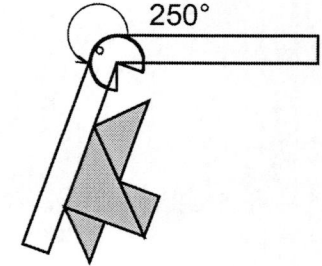

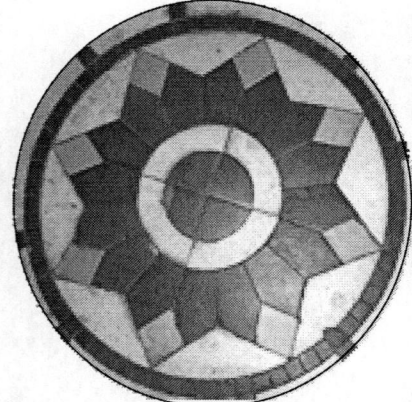

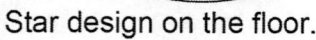

Star design on the floor.

> **Construct every line accurately. Don't guess.**

To set your ship aright, un-tape it, turn your goniometer upside down, re-tape it, and continue rotating as in the right figure above.

Worksheet. The first half of the worksheet asks you to construct the ship at various angles with your tools. You may find it helpful to set the ship model at the desired angle. Start your construction at the "×" and draw the first line at the correct angle. Measure only the line for the ship's bottom (3 cm); construct the other lines.

For the second half, build and rotate the model to the various angles before attempting the constructions. Measure only the 2.5 cm line.

Lesson 85

This lesson is similar to Lesson 114 in Level E/Gr 4.

Making Wheel Designs

GOAL 1. To make designs using rotational geometry

MATERIALS Worksheet 85
Goniometer
Scissors, tape
2 pieces of heavier paper 6 cm × 4 cm (Old business cards are ideal.)

ACTIVITIES *Rotating symmetry.* The pictures of wheels shown on this page have rotating symmetry. That is, the same design is rotated about a point. Computer design programs also have rotating symmetry where objects are not rotated, like the chairs on a Ferris wheel.

Preparing the template. For this lesson you will make a design for a wheel. The heavy paper will become a template for tracing your design. Start by making the height of the card equal to the radius of a circle on the worksheet. Next draw a design on the card. Then cut into the card and cut out the design, which is discarded. Next tape the two edges of the card back together. See the figures at the right. Lastly, draw marks at the top and bottom of your template.

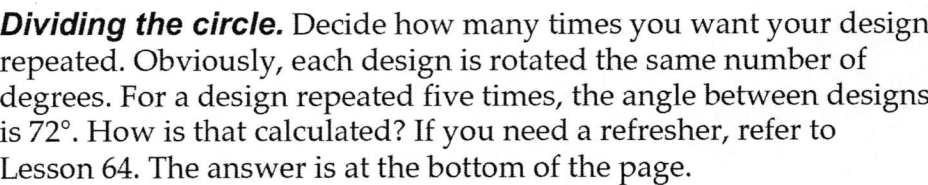

The greater number of repeats, the narrower your design must be to avoid overlapping.

Dividing the circle. Decide how many times you want your design repeated. Obviously, each design is rotated the same number of degrees. For a design repeated five times, the angle between designs is 72°. How is that calculated? If you need a refresher, refer to Lesson 64. The answer is at the bottom of the page.

Calculate the angle you need for your design. Then use your goniometer to make the tick marks on the circle. Align one mark of your template with the tick mark on the circle and the other mark on your template with the center of the circle. Trace your design. Repeat at each tick mark on the circle. See the figures below.

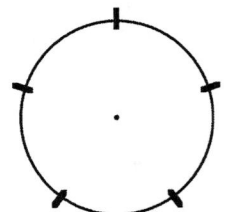 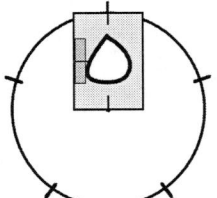 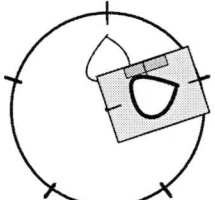 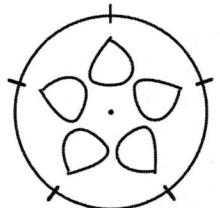

Worksheet. Make wheel designs in both circles, using a different number of rotations for the second circle. [Answer: 360 ÷ 5 = 72]

  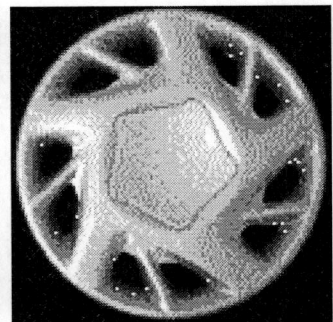 

Lesson 86

Identifying Reflections and Rotations

GOALS
1. To introduce the term *transformation*
2. To identify reflections and rotations

MATERIALS
Worksheet 86
A set of tangrams

ACTIVITIES
Transformations. In the last three lessons, you have reflected and rotated various objects and figures. The word that describes these changes is *transformation.* Reflections and rotations are congruent transformations. There is a third type of congruent transformation, called translation, which you will encounter in the next lesson.

In this lesson, you will identify whether a figure has been reflected horizontally or vertically or how much it has been rotated.

An example. See the numeral 2 and several reflections and rotations in the figures below. Try to decide how figures (a) to (e) relate to the first figure, the 2.

a.   b.   c.   d.   e.

A quick glance tells you that figure (a) is not a reflection. So it must be a rotation. You might imagine placing the original 2 on a goniometer and opening it 90°, which will give you figure (a). You can do the same thing with your left hand. Start with your hand flat like the base of the 2 and rotate about your wrist until it matches the base of figure (a). Notice that your hand moves through 90°.

Figure (b) also is not a reflection. Thinking of the goniometer or your hand will tell you it is a 180° rotation. You also know it's a 180 rotation because the figure is upside down.

Figure (c) is a reflection, actually a horizontal reflection. Or to be more formal, it is a reflection about a vertical line. Figure (d) is a vertical reflection; that is, a reflection about a horizontal line.

Figure (e) is a 270° rotation. Another to think about it is to think of rotating it clockwise 90°. Since it's the wrong direction, it's really –90°, which is the same as 270° because 360 – 90 = 270°. Your right hand can do the rotating.

Worksheet. On the worksheet are five rows having figures that are either reflected or rotated. In two cases there are two answers. At the bottom of the page is space for you to draw your own original figure and four reflections or rotations. Don't make them too complicated.

Rug design in hotel.

Lesson 87

Translations

GOALS 1. To learn the terms *translation* and the *image*
2. To realize the difference between *absolute* and *relative*
3. To create images through translations, using a coordinate system

MATERIALS Worksheet 87
Drawing board, T-square, 45 triangle

ACTIVITIES ***Translation.*** A mathematical *translation* means moving a figure from one point to another point without any reflections or rotations—nothing fancy, a plain move. Translation is the mathematical name for *slide.* The new figure is called the *image.*

In the figure below, point *A* is translated (6, –1) to point *A'* (read as *A prime*). The first number in the parentheses, 6, refers to the distance moved in the *x,* or horizontal direction.

The second number, –1, refers to the distance the point moved in the *y,* or vertical direction. Since the number is negative, the translation is 1 unit down, not up.

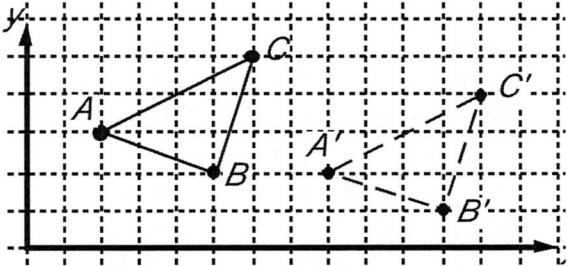

The image is named with the same letter or letters followed by a prime ('). So the translation of point *B* is *B'* and *C* is *C'.* The translation of ΔABC is ΔA'B'C'.

Absolute and relative. Point *A* is located at (2, 3) and point *A'* is located at (8, 2). These are their *absolute* positions. The translation (6, –1) is the *relative* distance between them, the amount of change. We make the same distinctions when we say the time is 1 o'clock (absolute) and math class lasts 1 hour (relative). Or the temperature is 20° (absolute) versus the temperature warming up 20° (relative).

Worksheet. On this worksheet you are to draw several images and name some coordinates. Problem 1 explores what happens with two translations.

Design on a door at Salt Lake City Airport.

G: © Activities for Learning, Inc. 2010

Lesson 88

Transformations

GOALS
1. To learn the term *transformation* and the three basic types of congruent transformations
2. To use transformations to construct figures

MATERIALS
Worksheet 88
A set of tangrams
Ruler
Drawing board, T-square, 45 triangle

ACTIVITIES

Transformations. A mathematical *transformation* involves creating an image from a geometrical object. The transformations you have worked with so far are congruent transformations. They include translations, reflections, and rotations.

They can be summarized in the following chart.

Common Name	Mathematical Term	Examples	Needed for Transformation
Slide	Translation		Distance & direction
Flip	Reflection		Line of reflection
Turn	Rotation		Point & angle

There are other transformations that change the size, which you will study later. They of course, are not congruent.

Worksheet. In Problems 1-3, you will be using all three transformations, translations, reflections, and rotations, to construct a cube, a ship and a chair. There are no grid lines, so measure carefully. Use your drawing tools wherever possible.

In problem 4, you will write the translations for the arrow.

Lesson 89

Double Reflections

GOALS 1. To investigate repeating the same reflection
2. To investigate two different reflections and a rotation

MATERIALS Worksheets 89-1, 89-2, 89-3
Drawing board, T-square, 30-60 triangle, 45 triangle
A set of tangrams

ACTIVITIES **Worksheet 1.** In the first worksheet, you are to do the reflection of a figure twice in the same direction. The result might be obvious to you. Then write your conclusion in a complete sentence. Do the worksheet now.

Worksheets 2 and 3. For problems 7, 8, 10, and 11, try to visualize the reflections and rotations before drawing them. If you need help, especially when rotating 180°, construct the figures with tangram pieces. Then move them as needed.

Do the remaining two worksheets before reading any farther.

Analysis. You can explain this results to someone else with a sheet of paper with text or a drawing. See the example below. The sheet of paper on the left is first flipped horizontally (middle figure) and then vertically (last figure). A different way to turn a piece of paper upside down!

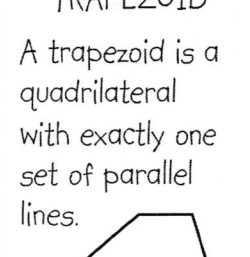  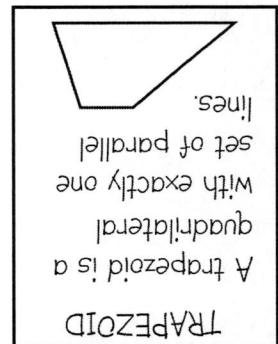

Lesson 90

Finding the Line of Reflection

GOALS 1. To find the line of reflection between an object and its image
2. To learn a new method of finding the midpoint (middle) of a line

MATERIALS Worksheet 90
Drawing board, T-square, 45 triangle
Ruler

ACTIVITIES *Line of reflection.* In previous lessons you were given the line of reflection. Now you will need to find the line between an object and its image.

Worksheet. The first problem is an investigation into finding a line of reflection. You are given a triangle and its image. They are shown on the right. Connect the corresponding points with a line segment (*A* to *A'*, *B* to *B'*, and *C* to *C'*). Use your drawing tools to find the perpendicular bisector of the three lines.

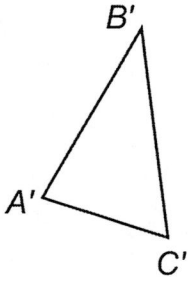 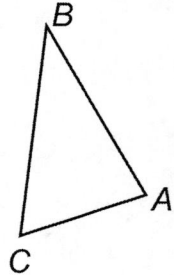

Do this problem now before reading any farther.

Analysis. You probably observed that the line of reflection bisects each line connecting the corresponding points. Is the line of reflection *perpendicular* to every line connecting corresponding points? The answer is at the bottom of the page.

Finding a midpoint. Here is a simple procedure for finding the middle of a line, using only a ruler First place the ruler near the two points, or end of the line. Second choose a whole number on the ruler that is near the center. Third, adjust the ruler so the fractional parts are the same on both sides of the points.

In the top figure below, 3 is near the center between the points. In the next figure, the ruler is moved to the left so there is .2 to the left of the 2 and .2 to the right of the 4. That means that 3 is 1.2 from each point, so the center is at 3, where a tick mark marks the spot.

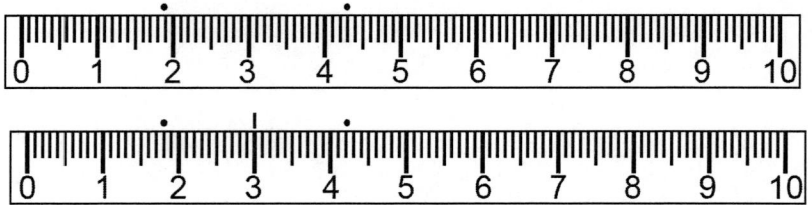

Check your work by visually connecting corresponding points. Then connect the centers.

Finding the line of reflection. To complete the worksheet, you will need to find seven lines of reflections. First find the midpoints of two corresponding points. Connect the tick marks, checking to be sure your line is perpendicular. That is the line of reflection. [Ans: yes]

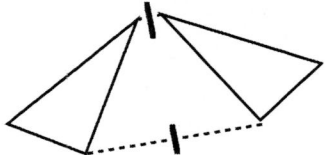

Lesson 91

Finding the Center of Rotation

GOALS
1. To find the center of rotation of an object and its image
2. To find the angle of rotation and to draw more images

MATERIALS
Worksheet 91
Drawing board, T-square, 45 triangle
mmArc Compass™
Goniometer

ACTIVITIES
Point of rotation. In the last lesson you found the line of reflection. Now you will find the point of rotation of an object and its image.

Worksheet. The first problem is an investigation into finding the point of rotation. The same triangle from the last lesson is rotated as shown on the right. Start by connecting the corresponding points. Then you need to find the perpendicular bisectors.

Perpendicular bisector. You can construct a perpendicular bisector with triangles (Lesson 32) or with a compass (Lesson 72).

Or you can use the following procedure with drawing tools. First, bisect the dotted line, as you learned in Lesson 51. See the left figure below. Then turn the T-square over, align the triangle to the dotted line. Without moving the T-square, moving the triangle in position to draw the perpendicular line. See the right figure below.

> *The largest tangram triangle makes a good tool for drawing the perpendicular bisector.*

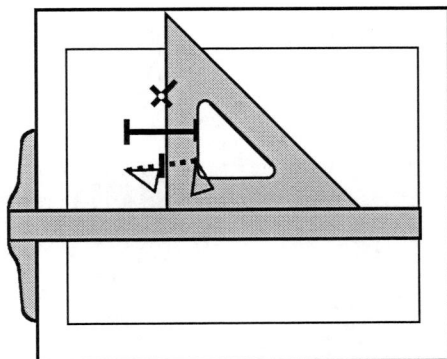 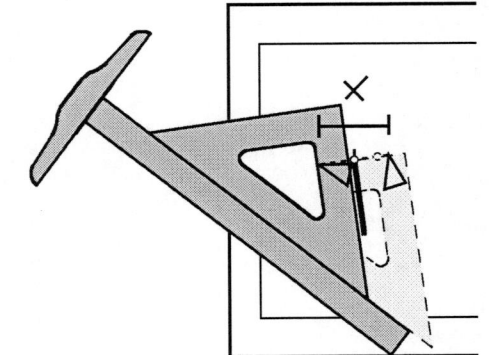

Worksheet. For Problem 1, draw the three lines connecting the correspond points and then draw their perpendicular bisectors. Extend the bisectors, if necessary, until they intersect. Do this now before reading any farther.

Analysis. Your three perpendicular bisectors should have intersected at the same point. Label it *P*. To see if point *P* is the same distance from *C* and *C'*, measure *CP* and *C'P*. Are they the same?

Worksheet. Complete the worksheet. Follow the instructions carefully. To draw the new images, first find the points on the circle and then draw the line segments.

Lesson 92

More Double Reflections

GOALS   1. To explore double reflections with different angles
2. To discover the location of the center of rotation

MATERIALS   Worksheets 92-1, 92-2
Drawing board, T-square, 45 triangle, 30-60 triangle
A set of tangrams (optional)
Goniometer

ACTIVITIES   ***Other double reflections.*** In Lesson 89, you discovered what happens when an object is reflected vertically and horizontally. The results give the same image as if the object had been rotated 180. This lesson explores reflecting through other angles.

Problem 1. On the worksheet, the first problem shows a square formed with tangram pieces. First reflect it around line *m*. Before you can use your drawing tools, you will need to measure to find a starting point. Repeat the reflection over line *n*.

Do this much before reading further.

Analysis. Compare your last image with the original object. Is it a reflection or a rotation? The answer is at the bottom of the page.

Measure the angle between the two lines of reflection, *m* and *n*. Consider line *m* rotating into line *n* to find whether the angle is positive or negative. Remember that if the rotation is counterclockwise, the rotation is positive. Write the angle in the table on the second worksheet.

Next find the center of rotation. Also find the angle of rotation and record it in the table.

Worksheet. Complete the worksheet. (Answer: rotation)

A building in Virginia Beach, Virginia.

Lesson 93

Angles of Incidence and Reflection

GOALS
1. To learn the term *angle of incidence*
2. To learn the term *angle of reflection*
3. To draw a series of angles of reflection

MATERIALS
Worksheet 93
Drawing board, T-square, 30-60 triangle, 45 triangle

ACTIVITIES
Angle of incidence. When a ray of light strikes a surface, the angle measured is between the ray and a line *perpendicular* to the surface at the point where the ray strikes the surface. This is called the *angle of incidence*. See the left figure below. *Incidence* means the point of contact.

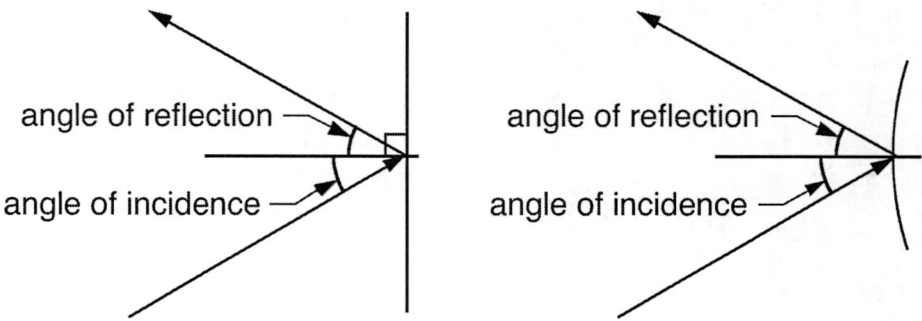

Angle of reflection. The ray of light will reflect from the surface at the *angle of reflection*. One of the principles of physics for light is that the angle of incidence is equal to the angle of reflection.

Even for curved surface, as in the right figure above, the principle is still true. The angles are always measured from the perpendicular to the surface.

Pool table game. For this game, a ball acts like a ray of light. That is, the angle of incidence equals the angle of reflection. In the real world, friction prevents them from being exactly equal. In the first figure below, the ball is aimed at 60°. In the second and third figures, the ball is reflected. The last figure shows the completed path before the ball drops into one of the cups in the corners.

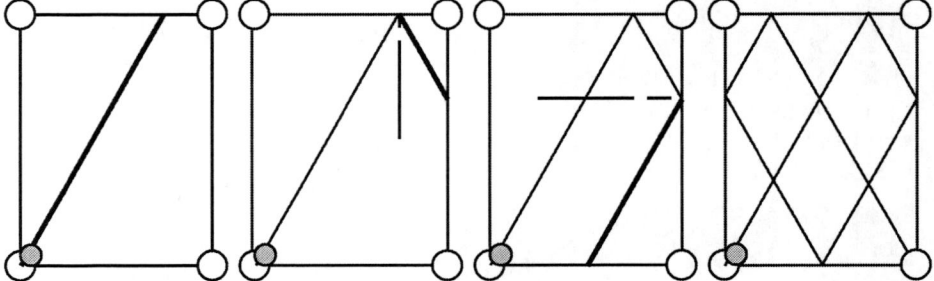

Worksheet. For the first three games, you can use your triangles. For problems 4 and 5, use the grids provided to draw your reflections. Note which pockets the ball drops into. Work carefully so your ball will land in a cup. For a variation of this game visit RightStartMath.com/geometry, and click the link for lesson 93.

Lesson 94

Lines of Symmetry

GOALS
1. To review *line of symmetry*
2. To compare a line of symmetry to a line of reflection
3. To find the lines of symmetry in a figure
4. To learn the terms *maximum* and *minimum*
5. To learn about the infinity symbol, "∞"

MATERIALS
Worksheets 94-1, 94-2
Drawing board, T-square, 30-60 triangle, 45 triangle
A set of tangrams

ACTIVITIES

Line of symmetry. You probably know what a *line of symmetry* is. It divides the figure into two parts with one part being the reflection, or mirror image, of the other part. A good way to check the line of symmetry is to fold the figure in half. If the two halves match, the fold line is the line of symmetry.

Line symmetry is very common in nature, in art, and in logos, which are symbols identifying a business or institution. See examples in the figures below. Note that the line of symmetry may be at any angle. Also a figure can have more than one line of symmetry.

From the garden of the Castle of Angers in Angers, France.

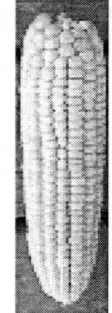  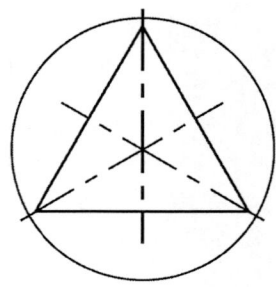

La Fleur de Lise, important symbol in France, in St. Malo, France.

Comparing to line of reflection. Reflections are a transformation, which means an object is transformed, or changed, into something else. Mathematicians use the word *map* to describe this transforming of an object according to some rule.

To think how a line of reflection is related to a line of symmetry, take a figure with a line of symmetry and remove half along a line of symmetry. Then reflect the remaining part about the line of symmetry, which becomes the line of reflection. Now you're back to the original object.

So, a line of symmetry is a line within a figure. A line of reflection is usually outside the figure and is the line the figure is flipped over.

Some face fun. Faces are almost symmetric. It is fun to make a picture of someone completely symmetric. To do this, first put the picture into computer software, either Adobe Photoshop or similar software. Then remove half. Copy the remaining half and reflect it horizontally. Lastly place the reflected along side the original half. See an example on the next page.

Design in the floor at the Minneapolis-St. Paul Airport.

From the botanical garden in Big Island, Hawaii

Building in Maryland.

A tulip from Monet's Gardens in Giverny, France.

To write the infinity symbol, "∞," start in the middle, draw a loop on one side and then the other side.

Original photo.

Photo with left side reflected.

Photo with right side reflected.

Worksheet. You can do the first 19 questions on the worksheet now or finish reading below before doing the worksheet.

Maximum and minimum. You may have seen the terms *maximum* and *minimum* on highway signs. The maximum speed is the highest speed a person may legally drive. The minimum speed is the lowest speed allowed on the road.

As an example for maximum and minimum number of lines of symmetry, let's take an isosceles triangle. See the left figure below, which shows the usual isosceles triangle. Clearly, it has one line of symmetry.

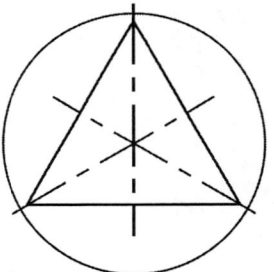

But the definition of a isosceles triangle says it has two equal sides. How about an equilateral triangle, shown above on the right with its three lines of symmetry. Doesn't it also have two equal sides? So, an equilateral triangle is a special case of an isosceles triangle. Therefore, the minimum number of lines of symmetry in an isosceles triangle is 1 and the maximum number is 3.

Infinity symbol. Infinity is not a number, but a concept. For example, how many numbers are there. We say the answer is infinite because whatever number you say, I can say one higher. There is no limit.

You will need the symbol for infinity, "∞" for the table in problem 21. Sometimes the symbol is called "lazy eight."

Problem 20. You may need to look up some of the definitions. Remember the difference between a polygon and a regular polygon.

If you have a partner, compare your answers in the table before looking at the solution key. Discuss any answers where you disagree.

Lesson 95

Rotation Symmetry

GOALS
1. To review or learn the concept of rotational symmetry
2. To learn the concept of *order of rotation symmetry*
3. To learn the concept of *point symmetry*

MATERIALS
Worksheet 95
Two sets of tangrams
Colored pencils

ACTIVITIES

Rotation symmetry. Just as line symmetry refers to a line of symmetry within an object, rotation symmetry refers to rotations within an object. If a figure can be rotated and is the same as before the rotation, it has rotation symmetry. In the left figure below, the abacus has rotational symmetry; you can turn it 180° and it will look exactly on the original abacus.

This lily demonstrates rotation symmetry

The leaves on a milkweed plant rotate so they receive as much sunlight as possible.

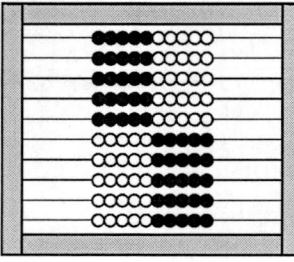
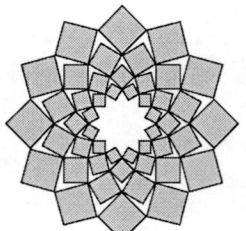

In the center figure above, an image of the design (actually a logo) can be rotated 30° and still fit exactly on the original. It can also be rotated 60°, 90°, 120°–that is, every multiple of 30° up to 360°, a total of 12 times. (We go around only once, so don't count anything past 360°.)

The car wheel above on the right can be rotated 72° (360 ÷ 5) and four more multiples of 72 and still look like it was originally. (Of course, you must ignore the black square around it.) The figure below shows the five counterclockwise rotations. There is a tick mark on top of the square to help you keep track as you observe the rotations.

The point of rotation is usually easy to find: the center of the figure.

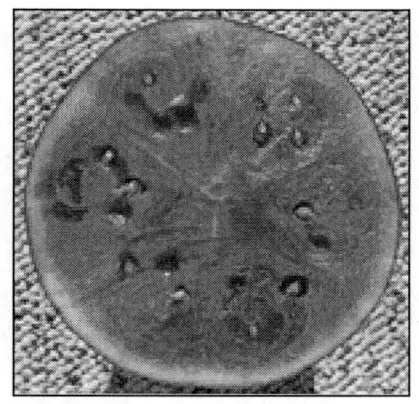

Watermelons also exhibit symmetry.

Sometimes you will see the term "2-fold" to indicate the order of 2, or "12-fold" for the order of 12.

Point symmetry. A special case of rotational symmetry is *point symmetry*. An easy way to check for point symmetry is to turn the object upside down. If it looks the same it has point symmetry.

To understand why it's called point symmetry, follow these steps. On a sheet of paper, make the figure shown below on the left with tangram pieces. Then turn the paper upside down to be sure it has point symmetry.

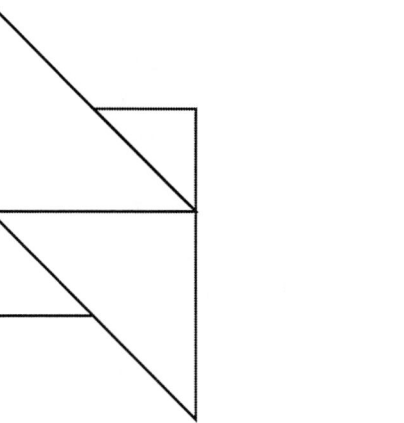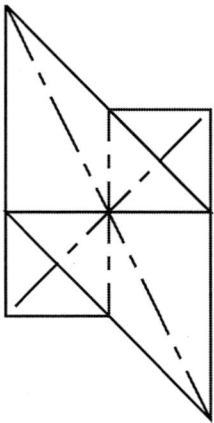

Next consider what happens if you connect the corresponding points. As you can see in the figure above on the right, the lines intersect at the center. The center also bisects each connecting line. It can be thought of as though each point is reflected through the center point. So, that's point symmetry.

Worksheet. The worksheet is a collection of problems applying symmetries.

Symmetry in logos. Collect a dozen or so logos from periodicals, the Internet, or on products. Analyze them for symmetry.

*Translating an object with point symmetry. Place an object with point symmetry, such as the parallelogram tangram piece on a sheet of paper. Translate it a short distance away. Do this with a second parallelogram tangram piece. See the figure below. Does the resulting figure have point symmetry? Try other translations.

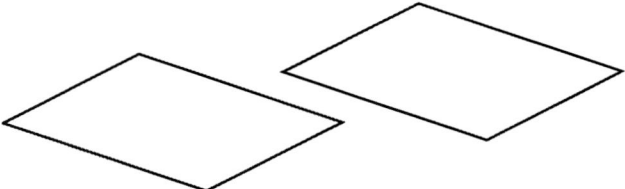

Lesson 96

Symmetry Connections

GOALS
1. To find the relationships between various types of symmetry in regular polygons
2. To discover some relationships between line symmetry, point symmetry, and rotation symmetry

MATERIALS
Worksheets 96-1, 96-2
Drawing board, T-square, 30-60 triangle, 45 triangle
A set of tangrams

ACTIVITIES
Worksheet 1. In the first worksheet you will be exploring line, point, and rotation symmetries in regular polygons. First you need to draw the lines of symmetry for six regular polygons. You can easily draw the lines of symmetry for four of the polygons. For the five-sided figure, you can draw the horizontal line of symmetry. To draw the other lines, you will need to bisect one side. Repeat for the seven-sided polygon.

After drawing the lines of symmetry, fill in the table. (The name of a 7-sided polygon is *heptagon.*) Then study the table and answer the question.

Worksheet 2. For the second worksheet, you needn't draw the lines of symmetry. Fill in the table, answering the three questions with either a Yes or No.

Be sure you understand Questions 5 and 6 before answering them.

This star was found in a museum in England.

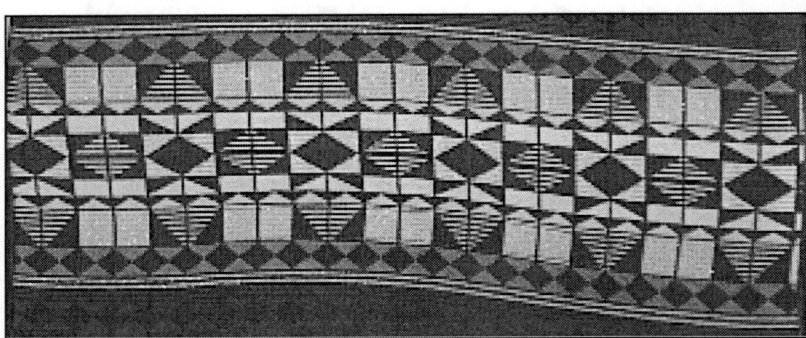
Design on Alaska Airlines passenger seats.

Lesson 97

Frieze Patterns

GOALS
1. To learn the terms, *frieze, cell,* and *tile*
2. To learn about the seven types of frieze patterns
3. To continue some frieze patterns
4. To construct an original frieze pattern

MATERIALS
Worksheets 97-1, 97-2
Drawing board, T-square, 45 triangle
A set of tangrams

ACTIVITIES

Coils for geothermal heating form a frieze pattern.

Not everyone uses the same names for the frieze patterns.

Design on a building in downtown Hutchinson, Minn.

Design on a building in Anchorage, Alaska.

Frieze patterns. A frieze (FREEZ) is a repeating pattern in one direction. You can see them, for example, as ornaments on buildings, as borders on walls, and on edges of rugs.

They are built from transformations you are familiar with. They include congruent translations, reflections, rotations, and combinations of these.

The smallest unit of a design is referred to as the *cell.* The cell for the following friezes will be the 30-60 triangle.

Translation. The most basic frieze of all is called a translation. It is merely a repeating along the same direction. Since all friezes use translations, that is not really a good name, but no one has a better name. See the figure below.

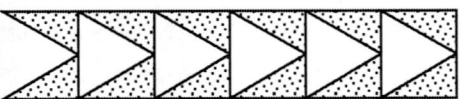 Translation

Vertical reflection. For the second frieze, flip the cell vertically and then translate them both. The cell and its reflection, that is, the unit being translated is called a *tile.* See the six tiles in the figure below.

Vertical Reflection

Horizontal reflection. For the third type of frieze, flip the cell horizontally and then translate the resulting tile. The figure below has three tiles.

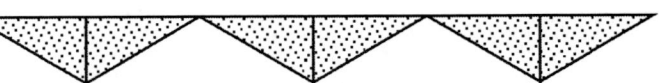 Horizontal Reflection

Vertical and horizontal reflection. For the fourth frieze, flip the cell vertically or horizontally and then flip them both the other way. Tile is composed of four cells. See the figure on the next page.

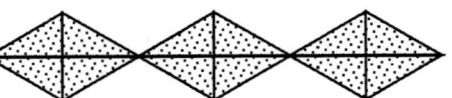 Vertical and Horizontal Reflection

Rotation. For the fifth frieze, rotate the cell. Then translate both. See the figure below with six translations.

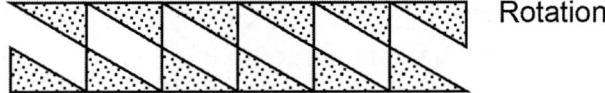

Rotation

Glide reflection. This is a different type of translation. It is a reflection and a translation and is called a glide reflection. Footprints a good example of a glide reflection.

Glide Reflection

Horizontal + glide reflection. The seventh and last type of frieze is the glide + glide reflection. It is shown below.

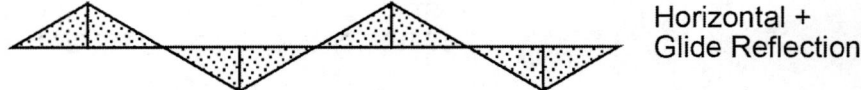

Horizontal + Glide Reflection

Identifying friezes. Mathematicians have determined that these seven types of friezes are the only ones possible. But it isn't always easy to tell which pattern is being used. Also, some variations are often used.

The frieze in figure (a) below is a translation, although it's in the vertical direction. In figure (b) the frieze pattern is vertical and horizontal reflection, but the vertical flip is below and not above as the example on the previous page. The rotation pattern in figure (c) has a different center of rotation from the example above.

The glide reflection in figure (d) uses a different line of reflection, moving the reflected images higher.

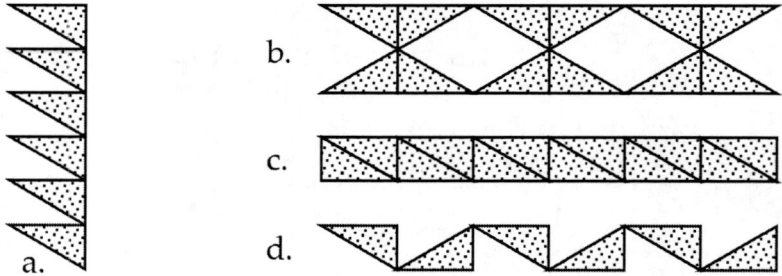

Worksheet. The worksheet asks you to draw six frieze patterns. The cell is two tangram pieces.

For the last problem, you choose your own cell and pattern. If you are working with others, try to determine which frieze patterns they used.

Lesson 98

Introduction to Tessellations

GOALS
1. To understand what a *tessellation* is
2. To complete a portion of Pentagon tiling #13
3. To find the internal angles in regular polygons

MATERIALS
Worksheets 98-1, 98-2
Drawing board, T-square, 30-60 triangle, goniometer

ACTIVITIES
Tessellations. A *tessellation* (TES-uh-LAY-shun) is a repeating pattern of distinct shapes that tiles, or covers, a surface without any gaps or overlaps. The word comes from the Latin word *tessella* meaning little squares, which were used to make mosaics.

Tessellations cover the whole area without end. Friezes, on the other hand, repeat a pattern in only one direction and often have gaps. Tiled floors are a good example of a tessellation. So are brick walls and patio surfaces. See some examples below.

Left picture: a floor design.

Right picture: a building at the University of Washington at Seattle, WA.

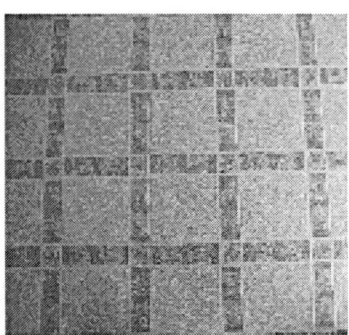

Types of tessellations. There are different types of tessellations. Some use only one shape; some use only regular polygons; and others use irregular shapes. Shown at the right is a tessellation, called the Pentagon tiling #13, which uses a special pentagon.

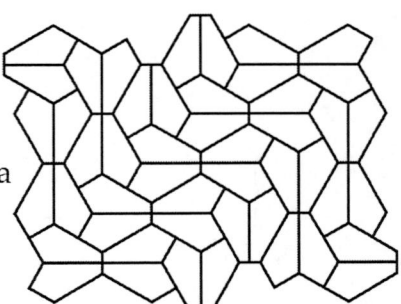

Pentagon tiling #13. The first five types of tessellating pentagons were found in 1918 by K. Reinhardt. By 1975, types 6-8 and 10 were discovered.

Marjorie Pick, from San Diego, California, discovered #9 and #11-13 in 1976-1977. She had not studied any math after high school. After reading an article by Martin Gardener about tiling pentagons in a copy of *Scientific American,* she became interested in the problem. You can read more about Marjorie Pick and the pentagon at http://www.rightstartgeometry.com. This site also shows a butterfly tessellation made from Pentagon tiling #13.

Pentagon tessellation #14 was discovered in 1985. No one knows if there are any more yet to be discovered. It is an unsolved mathematical problem.

Worksheets. Complete the first worksheet and answer the questions on the second worksheet.

Lesson 99

Two Pentagon Tessellations

GOALS
1. To construct two tessellations with the same pentagon
2. To draw the two tessellations
3. To learn the term *pure tessellation*

MATERIALS
Worksheets 99-1, 99-2
Scissors
Drawing board, T-square, 45 triangle

ACTIVITIES
Worksheets. Start by cutting out the 25 pentagons on the first worksheet. Discard the little triangles. There are two pentagons tessellations that you can make with these pentagons. Make them both and draw them on the second worksheet.

If you are truly stuck on finding the second tessellation, look below for a clue.

Pure tessellation. These two tessellations are examples of a *pure tessellation*, which is a tessellation using only one shape. The tessellation in the previous worksheet is another example.

Another pentagon tessellation. Shown below is Pentagon tiling #4. It is found in pavement designs in an ancient part of Cairo, Egypt. In this particular version, the pentagon is equilateral. For more examples of tessellations, see http://www.rightstartgeometry.com.

[Hint: a word in the title of Lesson 40]

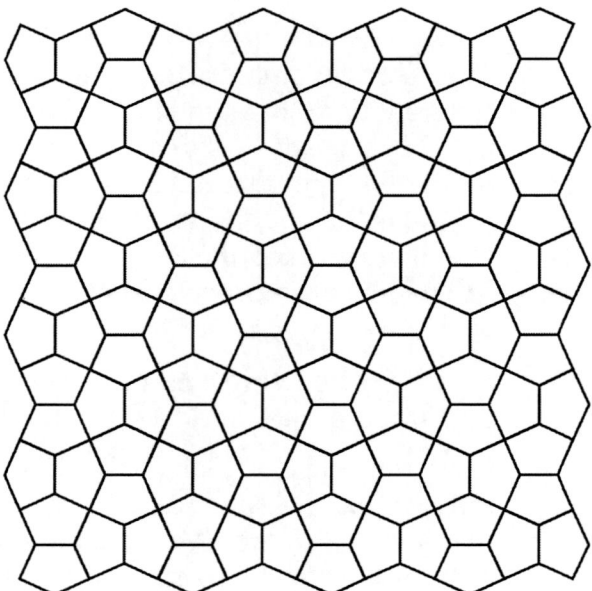

Lesson 100 **Regular Tessellations**

GOALS
1. To learn the names for the 9-, 10-, and 12-sided polygons
2. To find the internal angles in regular polygons
3. To find the regular tessellations
4. To draw the tessellations

MATERIALS
Worksheet 100
Drawing board, T-square, 30-60 triangle

ACTIVITIES
More polygon names. You have learned the polygon names up to octagon. Now you will need the names of three more, *nonagon* (NON-uh-gon), *decagon* (DEK-uh-gon), and *dodecagon* (DOE-DEK-uh-gon).

They are pictured below along with the number of sides. Notice the similarity of the words, *octagon, nonagon,* and *decagon* and the months, *October, November,* and *December.* In Latin, *oct-* means eight, while *non-* means nine, and *deca-* means ten. In the early Roman calendar, October was the eighth month.

> **How is a dodecagon like a clock?**

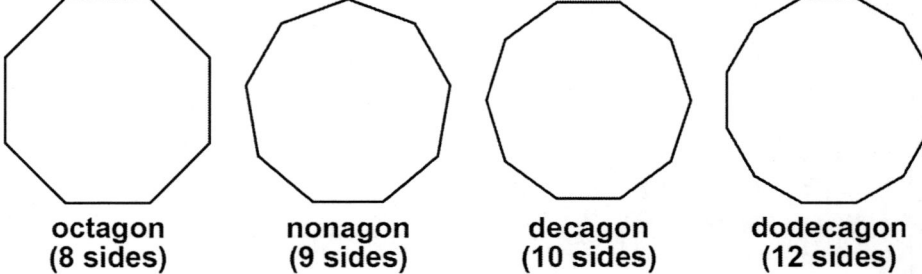

| octagon | nonagon | decagon | dodecagon |
| (8 sides) | (9 sides) | (10 sides) | (12 sides) |

In the word *dodecagon,* the *do-* (think of *duo* or *double)* means two. So a dodecagon has 2 and 10, or 12, sides.

Regular tessellations. A pure tessellation made with a regular polygon is called a *regular tessellation.*

Shown at the right are three pentagons arranged around a vertex. They do not form a tessellation because there is a gap.

Worksheet. The worksheet has a chart for you to complete. There are several ways to do the calculations. If you need a hint, refer to Lesson 44.

Use the information in the chart to find the gap in the regular pentagons and answer Question 2.

The answer to Question 3 is very important. Remember that to tessellate, there can be no overlaps or gaps. You know the sum of the angles around a point. So which three regular polygons tessellate?

Lastly, draw a regular tessellations for each regular polygon that tessellates. If you want more space, use the back of your worksheet.

Lesson 101 (2-3 days)

Semiregular Tessellations

GOALS
1. To learn the term *semiregular tessellation*
2. To find all the combinations to make semiregular tessellations
3. To learn the code of a tessellation

MATERIALS
Worksheets 101-2, 101-2, 101-3
Scissors
Drawing board, T-square, 30-60 triangle, 45 triangle, ruler
Colored pencils, colored pens

ACTIVITIES

Semiregular tessellations. As you learned from the previous lesson, the three regular tessellations aren't very exciting. But if you take more than one regular polygon, you can make some interesting designs. Of course, the regular polygons must fit around the vertex without overlaps or gaps.

A tessellation using more than one type of regular polygon is called a *semiregular tessellation.* The vertices must all have the same pattern. See an example at the right.

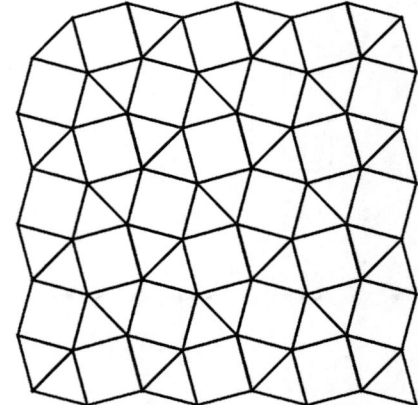

One of the eight semiregular tessellations. It is described as 3.3.4.3.4.

> **You will also need these polygon pieces for the next lesson.**

Worksheet 1. First cut out the regular polygons on the first worksheet as instructed. This will take about 8-10 minutes.

Worksheet 2. For the second worksheet, copy the interior angles from the previous worksheet. These five regular polygons are the only ones that will work to make semiregular tessellations.

Next figure out all the combinations that will total 360°. There are eight. One of the combinations will not work. Another combination makes two tessellations. This is easier to do if you get organized. Start with the largest angles first.

Construct each semiregular tessellation with the cutout pieces. Sketch the patterns on the worksheet, showing at least two vertices.

From a wall in Manhattan, New York

Naming semiregular tessellations. The angles are not used to describe a semiregular tessellation. Rather, a semiregular tessellation is identified by the number of sides of the polygons. Choose a vertex and start with the polygon with the lowest number of sides. Then name the number of sides of each polygon in order. Do it either counterclockwise or clockwise. So, for the above tessellation, its *code*, or numerical name, is 3.3.4.3.4. It could also be 3.4.3.3.4.

Worksheet 3. Draw and color a page of a semiregular tessellation.

*Colored lines. To color your lines, try this. First draw in pencil, but don't erase. Using a gel or Dry Erase pen, draw the horizontal lines starting at the top and continuing to the bottom. (Try out the pen and eraser on scratch paper first.) Wait until the ink is dry before drawing the remaining lines from side to side. When your work is completely dry, erase. All the pencil lines will disappear.

> **If you're finding pencil lines to be boring, here's how to add some color to your work.**

Lesson 102 (2 days)

Demiregular Tessellations

GOALS
1. To learn about *demiregular tessellations*
2. To learn the term *semi-pure tessellation*
3. To complete demiregular tessellations
4. To learn the codes of demiregular tessellations

MATERIALS
Worksheets 102-1, 102-2, 102-3
Drawing board, T-square, 30-60 triangle
Polygon shapes from the last lesson

ACTIVITIES
Who uses tessellations? No doubt you know that artists, architects, and interior designers use tessellations. They aren't the only ones. Many physicists, chemists, biologists, geologists, and engineers use tessellations because molecules and crystals form tessellations, usually in three dimensions.

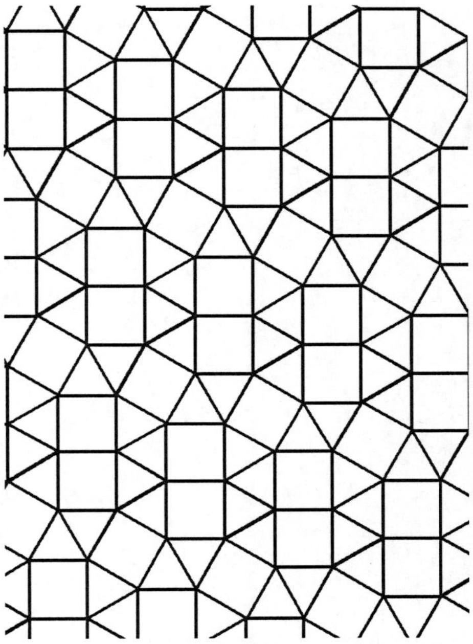

Demiregular tessellations.
Both semiregular and *demiregular tessellations* are constructed with regular polygons. While semi-regular tessellations have the same pattern at each vertex, demiregular tessellations have more than one vertex pattern. The prefix *demi* means approximately or about.

See the figure at the right.

Semi-pure tessellations. You recall that a pure tessellation was one made with only one shape. A tessellation made with more than one shape is called a *semi-pure* tessellation. That includes both semiregular and demiregular tessellations.

A demiregular tessellation. Its code is 3.3.3.4.4/3.3.4.3.4.

Numerical names for demiregular tessellations. Since a demiregular tessellation has at least two different vertex patterns, all vertices must be part of the code. Write the vertex patterns with a / between them.

The above tessellation is described as 3.3.3.4.4/3.3.4.3.4. Sometimes mathematicians tire of writing the repeating digits. They will use the shortcut $3^3.4^2/3^2.4.3.4$.

Worksheets 1-2. For Questions 5-6, use the polygon shapes to help you find the tessellations. If you need a hint, look at the end of the lesson. For Question 8, look for three different vertices.

Worksheets 3. Choose one of the demiregular tessellations on the worksheets or choose your own. Cover the page with your tessellation. You might copy it and give it to a younger person to color. Then color your original. [Hint: how can you get a 3.3.3.3.3.3 vertex inside a hexagon?]

Lesson 103

Pattern Units

GOALS   1. To learn a definition of *unit* and *pattern*
2. To practice finding the unit of a tessellation or pattern

MATERIALS   Worksheets 103
Drawing board, T-square, 30-60 triangle, 45 triangle, all optional
Ruler
Colored pens or pencils, optional

ACTIVITIES   ***Basic unit.*** A *cell* is the smallest repeating section of a frieze pattern (Lesson 97). For some unknown reason, when discussing tessellations, the smallest repeating section is called a *unit.* The statement often made is that "a pattern has three elements: a unit, repetition, and a system of organization." This also works in art.

Finding the unit. Translating the unit of a tessellation in two directions makes the entire design. First, translate the unit to form a row. Then translate the row to form the remaining rows.

Look at the figure on the right. Assume the pattern continues to infinity. What is the unit? Finding the unit sometimes takes trial and error. Obviously, the unit will include a dodecagon and at least one equilateral triangle.

Look at the figures below. Which one is the unit? To find out, mentally try making the design with each one. The answer is at the bottom of the page.

 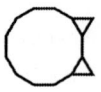 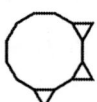 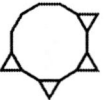 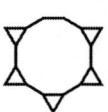 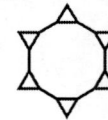

> *If you need to review translations, see Lesson 87.*

Finding the translations. To make a row for the tessellation above, measure where you want the second dodecagon to start. It will be (1 cm, 0) if you start in the upper left hand corner.

To find the translation to make the rows, draw a line as shown below in the left figure. Then measure the horizontal distance and the vertical distance as shown in the right figure. The translation becomes (.5 cm, –.87 cm). Of course, you could have started at one of the other three corners. The translations would have the same numerical value (same numbers), but different signs.

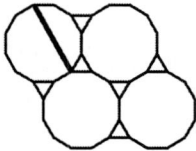   .87 cm —
.5 cm —

Worksheet. You need to find the units for all eight tessellations. Do this either by drawing over the unit or by coloring the unit. For Problems 1-3, you also need to find the translations needed to make the designs. [Answer: the secone one.]

Lesson 104

Dual Tessellations

GOALS
1. To learn what a dual of a tessellation is
2. To draw several duals
3. To find the relationship between a dual and its dual

MATERIALS
Worksheet 104
Drawing board, T-square, 30-60 triangle, 45 triangle

ACTIVITIES
Dual. The dual of a tessellation is a new tessellation formed by connecting the centers of the polygons around each vertex. Duals are constructed only with tessellations formed with regular polygons. The easiest way to understand the process is with an example.

In the left figure below is the semiregular tessellation with code 4.8.8. The center of each polygon is marked with an ×. The right figure shows the centers connected. Note that the new tessellation is not composed of regular polygons.

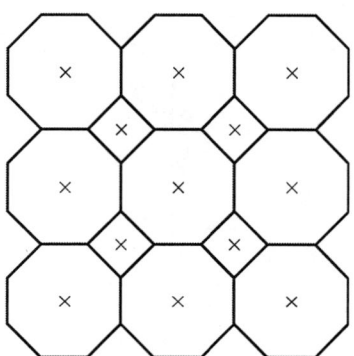 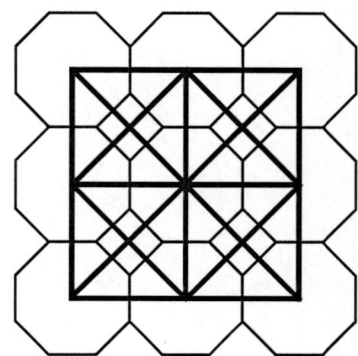

Finding centers of polygons. The center of a regular polygon is the same distance from all vertices and the same distance from all sides. Shown below are the constructions for finding the centers.

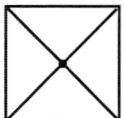 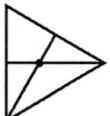 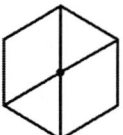 

Worksheet. You are to draw the duals for the tessellations on the worksheet. Be certain to draw lines connecting centers only when the polygons share a side, that is, when sides are touching. The polygons on the right don't share a side, so their centers cannot be connected.

An example of a demiregular tessellation and its dual.

  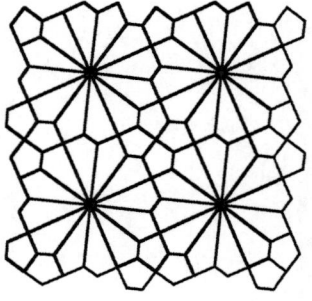

Tessellation   Tessellation and dual   Dual only

Lesson 105

Tartan Plaids

GOALS
1. To learn the terms, *tartan, plaid, warp, weft,* and *woof*
2. To understand the geometry of tartan plaids
3. To construct two plaids

MATERIALS
Worksheet 105
Drawing board, T-square, either triangle, ruler, scissors
Tracing paper, 9" × 6" (One sheet is included with your worksheets.)
Colored pens or pencils

ACTIVITIES
Tartan plaids. According to the Scots, who invented plaids hundreds of years ago, *tartan* is the pattern and *plaid* is the cloth. Today the word *plaid* can mean either the pattern or the cloth. The various Scottish clans each had their own plaid pattern.

There are at least 500 different designs. A few are shown below. See http://www.rightstartgeometry.com.

From the Elegant Stitches collection.

Weaving. Weaving is a process of making fabric from threads, or yarn. You probably did some simple weaving with paper strips when you were in the early grades. Shown below on the left is the simplest weaving pattern, called a *tabby*. The horizontal threads are called the *weft*, or *woof*, and the vertical threads are called the *warp*. Each weft thread goes over and under each warp thread in turn, creating a checkerboard pattern. This pattern is also used for making lattice pie crusts.

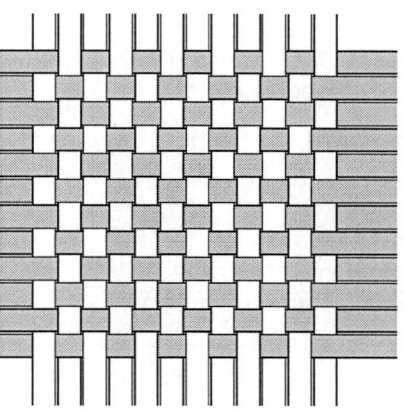
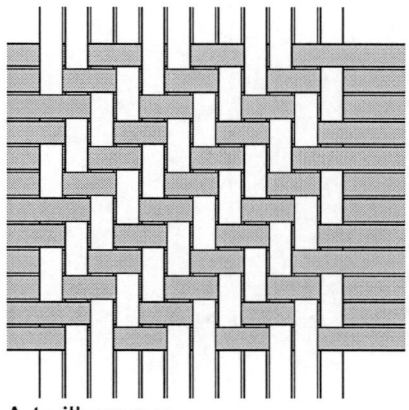

Plain weave called a tabby.   A twill weave.

The pattern above on the right is an example of a *twill* pattern. Notice that each weft thread goes under and over **two** warp threads. But instead of making the rows the same, they are offset by one. This produces the diagonal pattern as shown. Sometimes the number of threads the weft goes under is not the same as the threads it goes over.

Wool plaids are often woven as twill. Denim is a twill made with two colors, usually blue and white.

Colors in plaids. What is interesting is the color of the cloth after weaving with one color for the weft and another color for the warp. Our eyes see the overall color as a blend of the two. For example, blue and white results in light blue. It is similar to mixing paint.

Look at some plaid fabric or clothing. Notice that usually the same colors are used in both the weft and warp threads, but not necessarily in the same amounts. During the Middle Ages, the poor had plaids with one or two colors while the rich had more colors.

Worksheet. You will be constructing three plaid designs. Before you start, you will need to cut your tracing paper into rectangles.

First draw the lines for the design (one is done for you). Then tape the tracing paper over the design and color the rectangles representing the weft threads. Remove the tracing paper and color the rectangles for the warp threads on the worksheet. Finally replace the tracing paper.

Lesson 106

Tessellating Triangles

GOALS 1. To make tessellations using only triangles
2. To solve an area and perimeter problem

MATERIALS Worksheets 106-1, 106-2
Drawing board, T-square, 30-60 triangle
Colored pencils, optional

ACTIVITIES *Special triangles.* You already know that equilateral triangles tessellate. What about isosceles right triangles? A 45 triangle is half of a square and since squares tessellate, an isosceles right triangle must also tessellate.

Any triangle. Actually all triangles tessellate and in two different ways. An example of each type is shown below using a scalene obtuse triangle.

Both tessellations use rotations, but only the second one has reflections. The second row is a reflection of the first row.

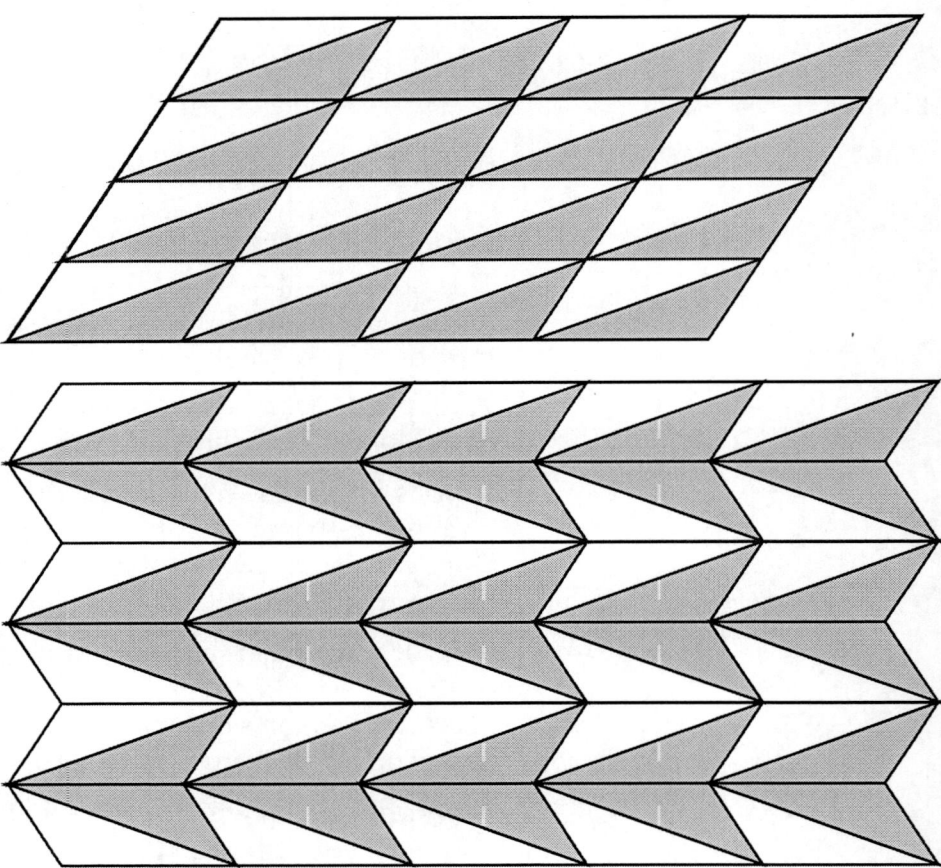

Worksheet 1. On the first worksheet, you are to draw both types of triangular tessellations, using a 30-60 triangle. Use your drawing tools to make it easier. Color the tessellations if you'd like. Sketch the pattern unit without using drawing tools.

Worksheet 2. If you need a hint for the second worksheet, look at your first tessellation on the other worksheet. You will need to divide several lines into fourths by dividing them in half twice. If you need help bisecting the lines, refer to Lesson 51.

Lesson 107

Tessellating Quadrilaterals

GOALS
1. To discover whether all quadrilaterals will tessellate
2. To make a tessellations using a given quadrilateral

MATERIALS
Worksheets 107-1, 107-2
Scissors
Drawing board, T-square, 30-60 triangle, 45 triangle

ACTIVITIES
Tessellating quadrilaterals. From Worksheet 1, cut out the four quadrilaterals on the left side. You are to arrange them so they will tessellate. Keep in mind two facts: the sum of the angles in a quadrilateral and the sum of angles around a vertex. The answers are at the bottom of the page. Arrange the four quadrilaterals now before reading farther.

Arranging the quadrilateral. Does your arrangement look like one of those below? These will not tessellate. Try again, but this time be sure all the sides line up. Do it now (if necessary) before reading further.

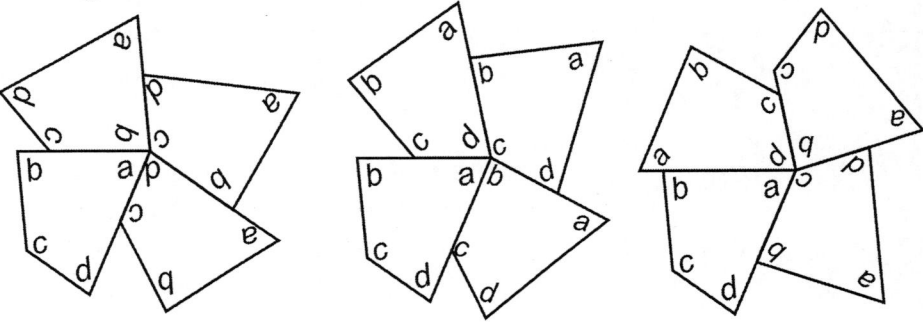

Now your arrangement should be similar to the arrangement below on the left. Can you imagine how the tessellation would look? The figure below on the right gives you a start.

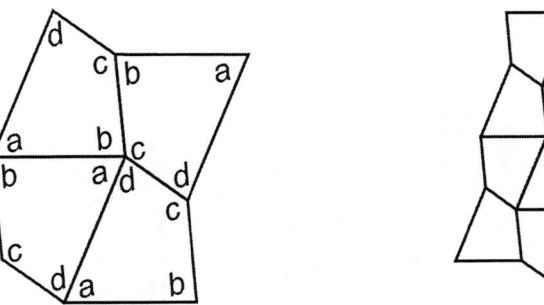

Worksheet 2. Cut out the remaining eight quadrilaterals on Worksheet 1. Arrange them into a tessellation based on what you just learned. This is your guide for Worksheet 2.

Draw the tessellation on Worksheet 2. One beginning quadrilateral is given. This quadrilateral was chosen because you can draw all the lines with your drawing tools. However, you will need to measure two of the lines in each quadrilateral. [Ans: both 360]

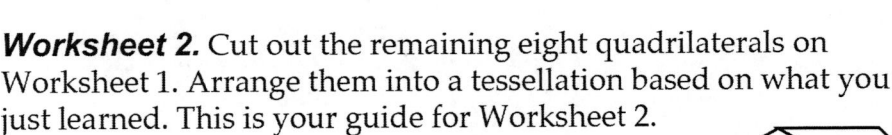

> Note that the quadrilaterals are NOT directly above each other.

Lesson 108

Escher Tessellations

GOALS
1. To learn a little about tessellations in history
2. To create some *Escher* type tessellations

MATERIALS
Worksheets 107-1, 108
A pen for making black, thick lines
Scissors and tape
Drawing board, T-square, 45 triangle
Colored pencils or pens

ACTIVITIES
History of tessellations. Tessellations have been found dating back to 4000 B.C. They have been incorporated into outsides of building, walls, floors, ceilings, rugs, and works of art.

Because the Islamic religion forbids representations of living creatures, Islamic cultures used geometry extensively. In Granada, Spain, the Moors created the beautiful Alhambra in the 1300s. The palace and mosque have many wonderful tessellations.

You can find many examples from various cultures at the following: http://www.rightstartgeometry.com.

Only in the last hundred years has the mathematics behind tessellations been studied extensively.

M. C. Escher. Maurits Escher, a Dutch artist, born in 1898, is famous for his tessellations of fish, horses, and other objects. In 1922, he visited the Alhambra, which greatly influenced his work. Escher also loved to draw illusions and impossible situations.

Below is an Escher drawing, including his original sketches. He started with a parallelogram tessellation and changed the sides. Then he added artistic details and alternated the colored of the birds, making every other one black. View more of his work at http://www.rightstartgeometry.com.

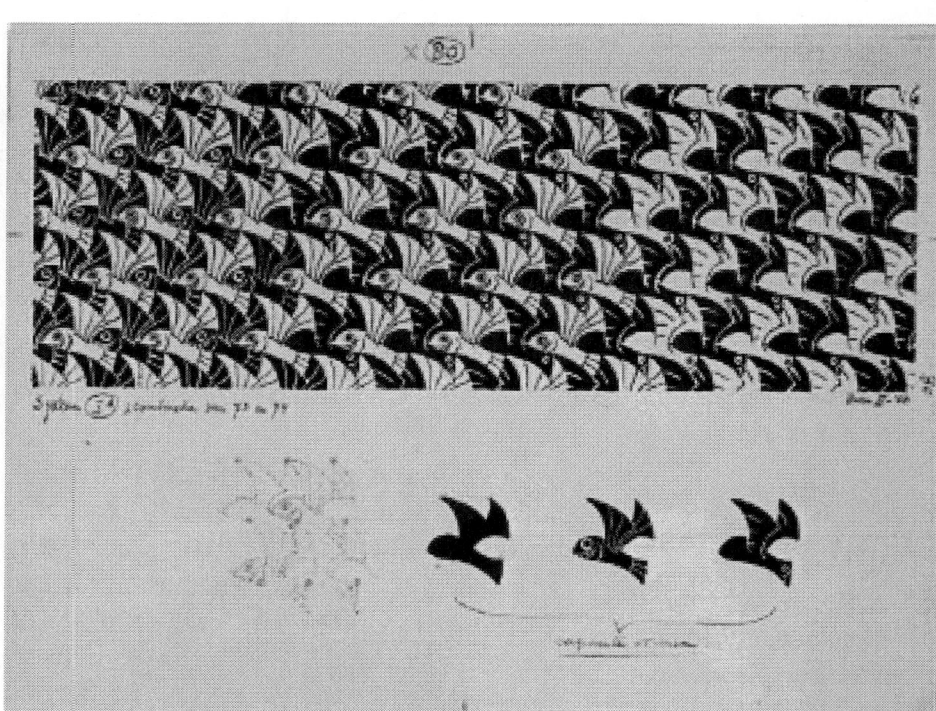

Drawing an Escher tessellation. The figures below show how the bird tessellation can be designed. Figure (a) shows the original parallelogram, which you know tessellates. Figure (b) shows a modification in the left side. This same change must be made on the right side as it is in Figure (c).

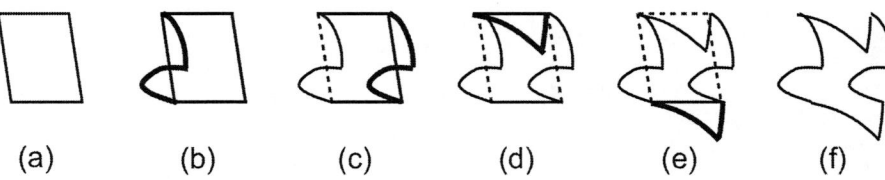

(a)   (b)   (c)   (d)   (e)   (f)

Next a part is removed from the upper edge of the parallelogram as in Figure (d) and added to the bottom in Figure (e). The final tessellating object is in Figure (f). It still needs details, such as eyes, beak, and feathers, placed so it knows which direction it is flying.

Worksheet. You are to make an Escher-type tessellation, starting with a rectangular tessellation. Instead of the entire tessellation, only the points for the vertices are given.

First experiment on the blank squares provided on Worksheet 1 from the previous lesson. Try various curves and lines, looking for something recognizable. Don't expect to find a design that's recognizable in two directions.

When you have found a suitable design, trace over it with the thick black pen. Then cut out around it. Tape your design to the drawing board and align the worksheet over the design and trace it. Repeat to make the tessellation.

The figure below shows a design copied once and the worksheet in position to copy the design again. Add details and color to your tessellation.

> **If your work is good, please send it to the author. It might be included in future editions.**

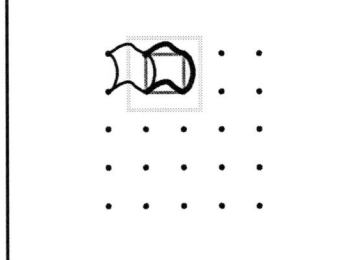

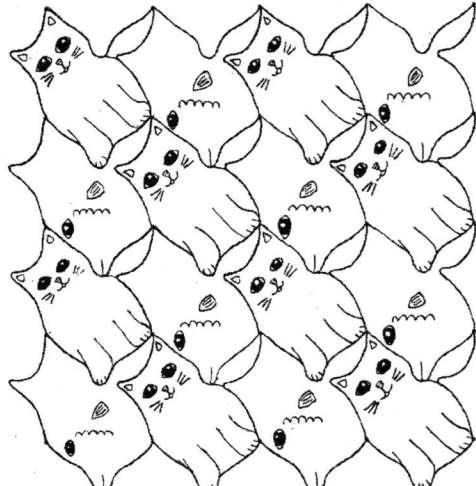

Tessellation by Renée S. Marais, 11, who says it started as angelfish, but some looked like cats, "And since cats eat fish, I call it 'The Food Chain.'"

Lesson 109

Tessellation Summary & Mondrian Art

GOALS 1. To review and summarize tessellations
2. To create some *Mondrian* art

MATERIALS Worksheets 109-1, 109-2
Drawing board, T-square, 30-60 triangle
Colored pens or pencils

ACTIVITIES *Mondrian art.* A Dutch artist, Pieter Mondrian, born in 1872, developed an art using rectangles of differing sizes. He used bright primary colors (red, blue, and yellow) and black and white. Many of the rectangles are left blank. An example is shown below. Although his art resembles tessellations, it does not have a repeating pattern. Learn more at http://www.rightstartgeometry.com.

Worksheet 1. For the last 11 or so lessons, you have been working on tessellations and related topics. This worksheet gives you a chance to understand the different types of tessellations and how they are related. Refer to previous lessons as you fill in the blanks.

Worksheet 2. Refer to Worksheet 1 to decide which categories the tessellations belong to.

Then on the lower half of the worksheet, draw two Mondrian designs. Fill in a few of your rectangles with primary colors. Carefully choose which rectangles to color, so the design is pleasing.

Lesson 110

Box Fractal

GOALS 1. To learn about *fractals* and the terms *iteration* and *self-similar*
2. To review or learn the meaning of *exponent*

MATERIALS Worksheets 110-1, 110-2
Drawing board, T-square, 45 triangle
Colored pens or pencils

ACTIVITIES ***Fractal.*** A *fractal* is a pattern, which repeats itself at smaller and smaller scales, resulting in an irregular shape. This definition will make more sense after you've seen examples of fractals.

Box fractal. Take a good look at the Box fractal shown below. Each *iteration* is one application of the instructions. Can you figure out how to go from the original square to iteration 1 or from iteration 1 to iteration 2? Think about it before reading farther.

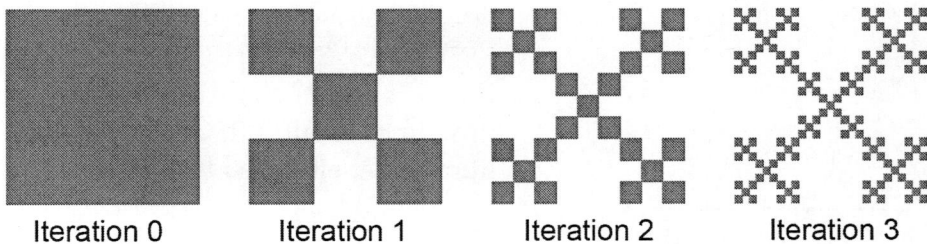

Iteration 0   Iteration 1   Iteration 2   Iteration 3

<u>Method 1.</u> One way to construct is to take the original square and shrink it so each side is one-third the length of the original. Then copy it and translate it to the four corners.

Repeat for the second iteration. Copy iteration 1, shrink, and translate. Continue forever or until you get tired or run out of space. The process is shown below.

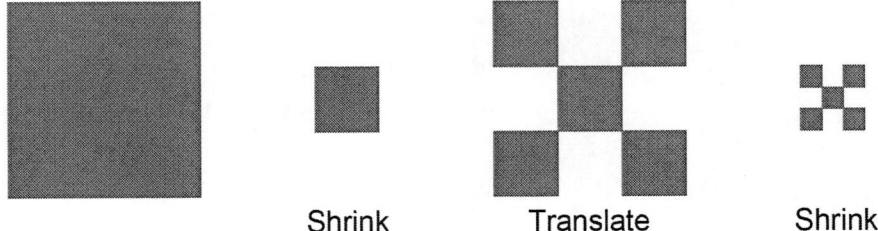

Shrink   Translate   Shrink

<u>Method 2.</u> Another method of making the Box fractal is to replace the sides of the original square with another figure. The new figure removes the middle third of a side and substitutes a square. See the figures below. The second figure shows the substitution at the

Replace ——————— with

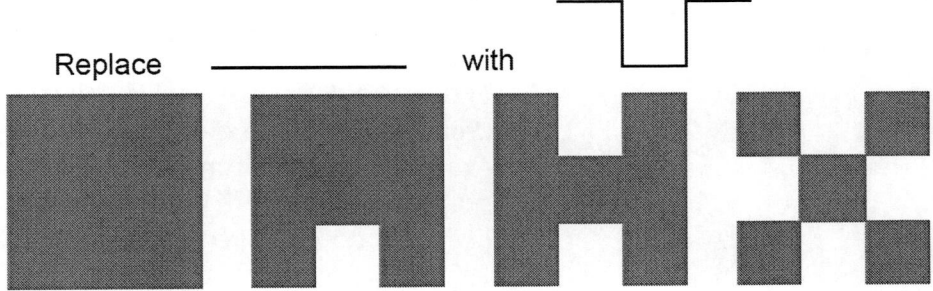

bottom and the third figure adds the substitution at the top. The last figure shows all four sides replaced.

Self-similar. When parts of a figure after magnification are similar to the whole figure, the figure is *self-similar*. Many objects in nature are approximately self-similar including clouds, coastlines, and mountains. Coastlines appear to be jagged from a plane, from the 20th floor of a tall building, from a car traveling nearby, and from a walk along the edge.

Worksheet 1. Draw iterations 1 and 2. The center squares give you starting points. To make the first iteration, you will need to make tick marks. Use the diagonal of the square as shown on the right. You have done this procedure before.

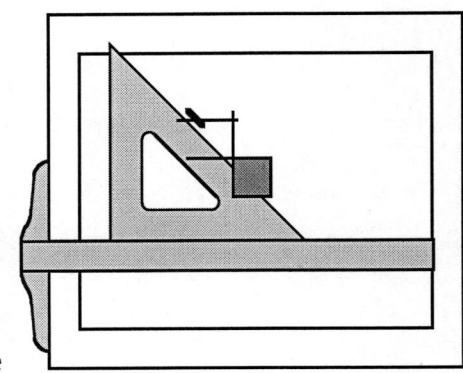

For iteration 2, you can use the little lines around it. Complete the iterations before reading farther.

If you know all about exponents, you can skim this paragraph.

Exponents. How could you find the number of squares in an iteration? Iteration 1 has five squares and iteration 2 has five times as many as iteration 1. Using multiplication, we write this as

$$5 \times 5 = 25$$

Using exponents, it is written as

$$5^2 = 25$$

The *2* is the *exponent* and tells how many times 5 is multiplied by itself. Exponents are a shortcut for writing repeated multiplication, just as multiplication is a shortcut for writing repeated addition.

For example,

$$4 \times 3 = 4 + 4 + 4$$
$$4^3 = 4 \times 4 \times 4$$

Worksheet 1 continued. For completing column 2 on the table, think of multiplying. Each iteration has how many times more squares than the previous iteration? In the third column, use exponents to write the same information as column 2.

In the fourth column, calculate the first two rows as usual. Then think about the ratio of the perimeter of each new iteration to the previous iteration. Use that to find the other perimeters

Do the same for the last column. Find the first few areas. Then use the ratio for the areas for adjacent (next to) iterations.

Worksheet 2. This worksheet asks you to graph the perimeter and area. For the perimeter graph the perimeter for each iteration. For example, for iteration 1, the perimeter is 18. Estimate where the point would be; it is lightly drawn. Plot the remaining points and draw a smooth curve connecting the points.

For the area graph, first decide on a scale so all your points will fit.

Lesson 111   **Sierpinski Triangle**

GOALS   1. To construct the fractal known as *Sierpinski Triangle*
2. To review a fraction of a fraction and percents

MATERIALS   Worksheets 111-1, 111-2
Drawing board, T-square, 30-60 triangle
Colored pens or pencils

ACTIVITIES   ***History of geometry.*** Euclid is the name usually associated with basic geometry. He lived in Alexandria, Egypt, around 300 B.C. People, of course, were familiar with lines, polygons, circles, and other geometrical objects before then, but Euclid organized the study of geometry with a system of reasoning and logic. Euclid wrote the book, *The Elements*, which remained a textbook for almost 2000 years.

History of fractals. Some of the basic ideas of fractals were known in the late 1800s. *Fractal* wasn't even a word until 1975 when Benoit Mandelbrot wrote *Les objets fractals.* In 1988 he wrote *The Fractal Geometry of Nature.* Mandelbrot organized the study of fractals and is considered to be the founder of the subject.

Sierpinski Triangle. The *Sierpinski Triangle* is a very famous fractal. Polish Waclaw Sierpinski (1882-1969) generated in 1915. Sometimes it is called the *Sierpinski Sieve* or *Sierpinski Gasket.* See the figures below.

Iteration 0

Iteration 1

Iteration 2

The fractal can be constructed by dividing an equilateral triangle into fourths and removing the middle triangle. Then the remaining three triangles are divided the same way. What fraction of iteration 1 is shaded? The answer is at the bottom of the page.

Fraction of fractal shaded. What fraction of iteration 2 is shaded? Well, three fourths of each shaded triangle in iteration 1 is shaded. So, the fraction shaded in iteration 2 must be three fourths of three fourths, which is nine sixteenths. Is this more or less than one half? The answer is at the bottom of the page.

Which arithmetic operation do you use to find a fraction of a fraction? The answer is at the bottom of the page.

Iteration 4

Worksheets. The first thing the worksheet asks you to do is to construct three iterations of the Sierpinski Triangle. Some hints are given at the bottom; read them if you need help. Use exponents for the first column of the table. [Ans. 3/4; more; multiplication; *(a)* find the center of the base line, *(b)* draw from the center to each side using your 60 angle.]

Lesson 112

Koch Snowflake

GOALS
1. To realize fractals exist in nature
2. To learn about some uses of fractals
3. To learn about the *Koch Snowflake* fractal

MATERIALS
Worksheets 112-1, 112-2
Drawing board, T-square, 30-60 triangle
Colored pens or pencils

ACTIVITIES
Fractals in nature. There are other types of fractals. A canopy fractal, so named because it looks like treetops, is shown below. The canopy itself looks the most like other fractals. This fractal is made by starting with a "Y," then copying, shrinking, and turning the Y (45° in this example) so the stem of the new Y aligns with a branch of the previous Y. See more at http://www.rightstartgeometry.com.

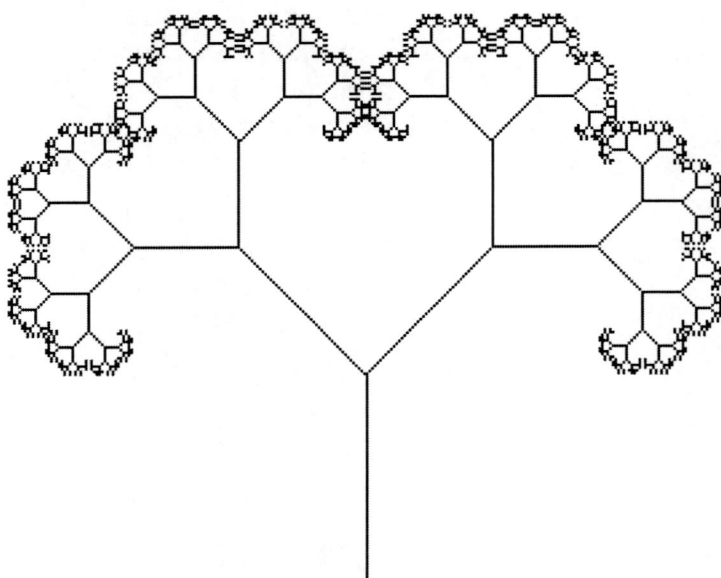

Canopy fractals even exist in your body. Consider your circulatory system with the aorta, which leads to arteries, which leads to blood vessels, and so on down to capillaries. And the capillaries are close to each other, which is another characteristic of fractals.

Your lungs along with your trachea, bronchial tubes, smaller tubes, and aveoli, are another fractal system.

Uses of fractals. Fractals are used extensively in computers. They save vast amounts of storage space and greatly reduce download times. You may have seen them in films, such as *Jurassic Park* and *Star Trek II*. The landscapes were designed with fractals.

Michael Barnsley designed the well-known Barnsley fern, shown at the right. Notice that each leaf has the same shape as the entire fern. The fern's stem is a leaf that has been squashed to the thickness of a line.

Barnsley Fern

Koch Snowflake. The Koch Snowflake is another very famous fractal. It was first published in 1904. The basic pattern is an equilateral triangle. Several iterations are shown below.

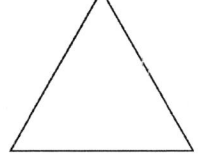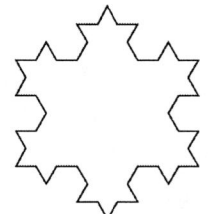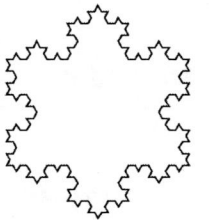

Worksheet 1. Here you will need to discover the basic pattern for the Koch Snowflake and construct iteration 2. Triangle grid paper is provided to eliminate measuring. Use your 30-60 triangle and T-square to make drawing easier.

Perimeter. Rather than measure the perimeter in centimeters, use the edge of a small equilateral triangle on the grid paper. Figure out some shortcuts so you don't have to count every little line. After you find the perimeter for the first few iterations, think about finding the other perimeters using a ratio. Keep in mind that each new side is four thirds as long as the original side.

Area. Use the small equilateral triangles on the grid as the unit for area. The outlines for iterations 0 and 1 are in dotted form on the grid.

Worksheet 2. Find the perimeters and areas to complete the table. Then graph the results. You will notice something very remarkable. Complete the worksheet before reading farther.

Analysis. As you probably guessed the perimeter becomes infinite. This is always true with fractals. Compare the area graph of the Koch Snowflake with the area of the Box Fractal. The area of the Snowflake "levels off." That means it doesn't keep getting larger as the area of the Box Fractal did. It is *finite* (FIGH-nite), the opposite of *infinite.*

Advanced math shows that the Snowflake's area is 129.6. That means if you spent the rest of your life making more and more iterations, the area would never go past 126.9.

132

Lesson 113

Cotter Tens Fractal

GOALS
1. To construct the Cotter Tens Fractal
2. To learn about naming large numbers
3. To learn about multiplying exponents

MATERIALS
Worksheets 113-1, 113-2
Drawing board, T-square, 30-60 triangle
Colored pens or pencils

ACTIVITIES
Cotter Tens Fractal. Although the Cotter Tens (see the figures below) looks at first like the Sierpinski Triangle, there are important differences. The Cotter Tens is based on ten shaded triangles instead of three. Regard the unshaded areas in both fractals as holes, or empty spaces. Also, look down the centerline. The Tens Fractal does not exhibit triangles increasing in size, like the Sierpinski Triangle.

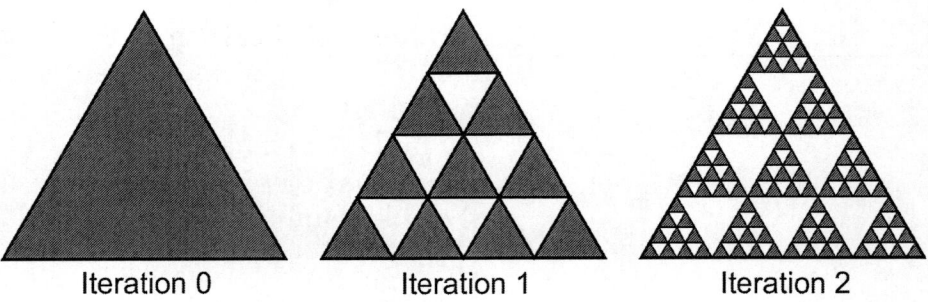

Iteration 0   Iteration 1   Iteration 2

Worksheet 1. For the first worksheet, you are to draw the first and second iterations of the Cotter Tens Fractal. Use the same techniques that you did for the Sierpinski Triangle.

Worksheet 2. Fill in the table. For column 2, be sure to write the number with exponents and with numerals. Carefully observe the patterns in your table to help you answer the questions. Do the two worksheets before reading any farther.

More fractals. Jason Padgett has a website where he displays fractals and other geometry he has drawn using pencil, ruler, and compass. See them at http://www.rightstartgeometry.com.

Millions. Our words for large numbers are based on millions. The *mil* part of *million* means thousand. For example, a mile is 1000 strides. A *stride* is the distance your foot travels when walking until it touches the ground again. See the figure below.

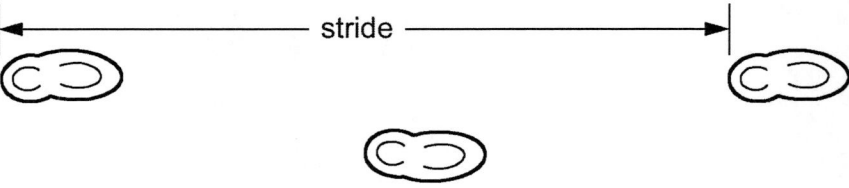

stride

G: © Activities for Learning, Inc. 2010

Billions and beyond. In the U.S. and Canada, a billion is a thousand millions. A trillion is a thousand billions and a quadrillion is a thousand trillions.

Reading large numbers. Here is a simple way to read large numbers, starting at the left. For example, look at the second row in the chart below. Read the part of the number up to the first comma, which is 1. Then place a pencil or your finger after the first comma and scan to the right to see the number of remaining commas. [1] Since there's one comma, say "million."

Example	Number of Commas	Name	Word Reminder
1,000	0	thousand	
1,111,111	1	million	
2,222,222,222	2	billion	*bicycle, bisect*
3,333,333,333,333	3	trillion	*triangle*
4,444,444,444,444,444	4	quadrillion	*quadrilateral*

Next try the fourth row. Read 3 and scan for the number of commas. Since there are 3, the chart says to say "trillion."

Now try a larger number, 123,456,789,123,456,789. Look at the chart below. Practice until you find it is easy.

Number	Notice	Read
1 2 3, 4 5 6, 7 8 9, 1 2 3, 4 5 6, 7 8 9	*4 commas*	123 quadrillion
1 2 3, 4 5 6, 7 8 9, 1 2 3, 4 5 6, 7 8 9	*3 commas*	456 trillion
1 2 3, 4 5 6, 7 8 9, 1 2 3, 4 5 6, 7 8 9	*2 commas*	789 billion
1 2 3, 4 5 6, 7 8 9, 1 2 3, 4 5 6, 7 8 9	*1 comma*	123 million
1 2 3, 4 5 6, 7 8 9, 1 2 3, 4 5 6, 7 8 9	*0 comma 1 group*	456 thousand
1 2 3, 4 5 6, 7 8 9, 1 2 3, 4 5 6, 7 8 9	*0 comma 0 group*	789

Lesson 114

Similar Triangles

GOALS
1. To expand ways of writing ratios
2. To review the mathematical definition of *similar*
3. To discover the properties of *similar triangles*

MATERIALS
Worksheets 114-1, 114-2
Ruler
Goniometer
Drawing board, T-square, 30-60 triangle, 45 triangle

ACTIVITIES

Ratio review. A ratio is the division of a number by another number. It expresses how many times larger or smaller a quantity is compared to another quantity. Notice that neither addition nor subtraction is ever part of ratios.

Writing ratios was discussed earlier in Lesson 23. In the figure on the right, the ratio of $\overline{LI}$ to $\overline{NE}$ can be written as 3 to 4, or 3:4, or ¾. Ratios look like fractions and act like fractions. They are fractions!

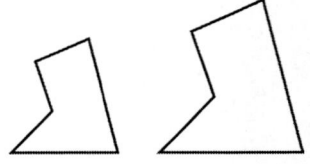

Since ratios are fractions, you can also write the ratio of the two lines as .75 or even 75%. This means that line *LI* is .75 or 75% as long as line *NE*.

Similar. As discussed in Lesson 49, *similar* means almost the same in everyday language. In mathematics *similar* figures are identical in shape, but may be shrunk or blown up. The triangles in the figure below are similar and the pentagons are similar.

Worksheets. The first problem asks you to find the ratios of the corresponding sides. Corresponding means matching. You also need to compare the angles.

For Problem 7, use your drawing tools. In Problem 9 do not confuse one-fourth more with one fourth. For example, if you have $8 and you receive one fourth more, you will have two more dollars, or ten altogether. Do the worksheet now before reading any farther.

Summary. In Lesson 49 you learned that two triangles with the same angles need not be the same size. In this lesson you learned such triangles are called similar triangles. You also learned that the ratios of their corresponding sides are equal. A short summary is: two triangles are similar if two corresponding angles are equal.

You need to know. ➤

Lesson 115

Fractions on the Multiplication Table

GOALS 1. To discover fractions on the multiplication table
2. To learn the term *proportion*

MATERIALS Worksheet 115

ACTIVITIES ***Fractions on the multiplication table.*** Did you know that you could find fractions on the multiplication table? For example, the fractions $\frac{1}{2}$, $\frac{3}{6}$, and $\frac{6}{12}$ are shown outlined with thicker squares in the left figure below.

1	2	3	4	5	6	7	8	9	10
2	4	6	8	10	12	14	16	18	20
3	6	9	12	15	18	21	24	27	30
4	8	12	16	20	24	28	32	36	40
5	10	15	20	25	30	35	40	45	50

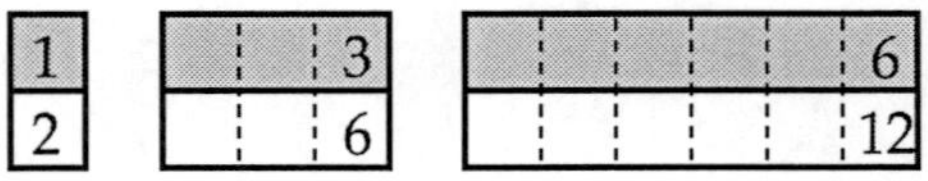

The equivalent squares are shown on the right figure above. Are the three fractions equal? Note that 1 is half of 2, 3 is half of 6, and 6 is half of 12. Do you see the other seven fractions in the first two rows of the multiplication table? Are they all equal to one-half?

Try another set of fractions, such as $\frac{6}{18}$ and $\frac{10}{30}$. See the figures below. Are they equal?

1	2	3	4	5	6	7	8	9	10
2	4	6	8	10	12	14	16	18	20
3	6	9	12	15	18	21	24	27	30
4	8	12	16	20	24	28	32	36	40
5	10	15	20	25	30	35	40	45	50
6	12	18	24	30	36	42	48	54	60
7	14	21	28	35	42	49	56	63	70
8	16	24	32	40	48	56	64	72	80
9	18	27	36	45	54	63	72	81	90
10	20	30	40	50	60	70	80	90	100

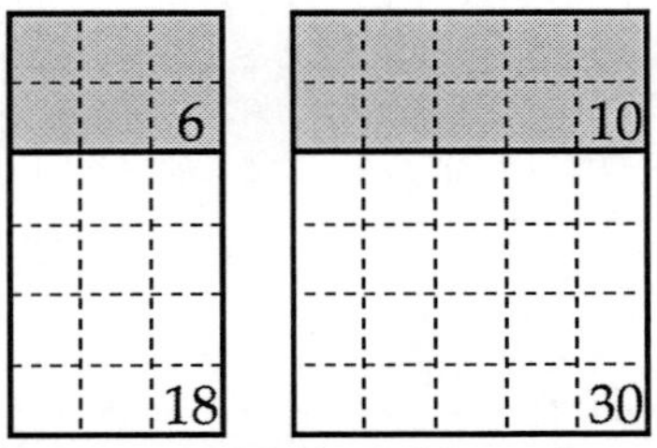

6 is one third of 18 and
10 is one third of 30.

Seeing more fractions. Look again at the squares at the top of the page. There is another set of fractions that can be written. Change

$$\frac{1}{2} = \frac{3}{6} \quad \text{to} \quad \frac{1}{3} = \frac{2}{6}.$$

Try it also for the figure above.
Change the equation from

$$\frac{6}{18} = \frac{10}{30} \quad \text{to} \quad \frac{6}{10} = \frac{18}{30}.$$

Observe what happens when the two new fractions are added to the multiplication table. (They are shown shaded.) Notice the symmetry; for example, the 6-square is reflected about the diagonal line. The other squares, 10, 18, and 30, are also reflected.

1	2	3	4	5	6	7	8	9	10
2	4	6	8	10	12	14	16	18	20
3	6	9	12	15	18	21	24	27	30
4	8	12	16	20	24	28	32	36	40
5	10	15	20	25	30	35	40	45	50
6	12	18	24	30	36	42	48	54	60
7	14	21	28	35	42	49	56	63	70
8	16	24	32	40	48	56	64	72	80
9	18	27	36	45	54	63	72	81	90
10	20	30	40	50	60	70	80	90	100

Proportion. When two ratios are equal, the result is called a *proportion*. The four equations in the previous section were all proportions.

Solving proportions on the multiplication table. It sounds crazy, but it's possible to use the multiplication table to solve proportions like

$$\frac{9}{24} = \frac{a}{56}.$$

First find a common (same) column for the 9 and 24. See the left figure below. Be sure the 56 fits in the same row as the 24. Mark the three numbers. Then find the fourth corner of the "rectangle" and you have the answer, $a = 21$. See the right figure below.

1	2	3	4	5	6	7	8	9	10
2	4	6	8	10	12	14	16	18	20
3	6	9	12	15	18	21	24	27	30
4	8	12	16	20	24	28	32	36	40
5	10	15	20	25	30	35	40	45	50
6	12	18	24	30	36	42	48	54	60
7	14	21	28	35	42	49	56	63	70
8	16	24	32	40	48	56	64	72	80
9	18	27	36	45	54	63	72	81	90
10	20	30	40	50	60	70	80	90	100

1	2	3	4	5	6	7	8	9	10
2	4	6	8	10	12	14	16	18	20
3	6	9	12	15	18	21	24	27	30
4	8	12	16	20	24	28	32	36	40
5	10	15	20	25	30	35	40	45	50
6	12	18	24	30	36	42	48	54	60
7	14	21	28	35	42	49	56	63	70
8	16	24	32	40	48	56	64	72	80
9	18	27	36	45	54	63	72	81	90
10	20	30	40	50	60	70	80	90	100

You could also have found a row, or column, for 24 and 56 and found a place for the 9. The last corner will be 21 in any case.

Worksheet. Do Problems 1-4 before reading farther.

Setting up a proportion. To find h in the smaller triangle on the right, you can set up a proportion,

$$\frac{h}{30} = \frac{80}{40} \quad \text{or} \quad \frac{80}{h} = \frac{40}{30},$$

or six other variations. You could solve it by using the multiplication table. Or simply notice that the hypotenuse is twice the length of the base line, so it must be 60.

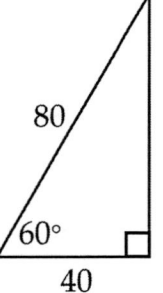
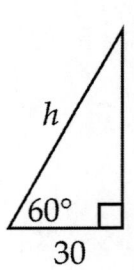

Worksheet. Complete the worksheet. To find e in Problem 5, remember the Pythagorean theorem.

Lesson 116

Cross Multiplying on the Multiplication Table

GOALS 1. To understand *cross-multiplying*
2. To apply cross-multiplying

MATERIALS Worksheets 116-1, 116-2

Cross multiplying. There is something else very interesting concerning fractions and ratios on the multiplication table. Refer to the first example from the previous lesson. Multiply the numerator of a fraction by the denominator of the other fraction, $1 \times 6 = 3 \times 2$.

You can see that they are equal by studying the rectangles in the middle figure below. The 6-rectangle is 3×2, making the product of the numerator of one fraction and the denominator of the other fraction equal: $1 \times 6 = 3 \times 2$. This forms a relationship called *cross multiplying*, indicated in the third figure below.

1	2	3	4	5	6	7	8	9	10
2	4	6	8	10	12	14	16	18	20
3	6	9	12	15	18	21	24	27	30
4	8	12	16	20	24	28	32	36	40
5	10	15	20	25	30	35	40	45	50

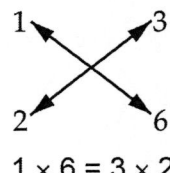

$1 \times 6 = 3 \times 2$

The second example is from the previous lesson: 6:18::10:30. The middle figure below shows the rectangles. Notice what happens when you cross multiply, 6×30 and 10×18. They are both equal to 180. Look at the second figure below and see that

1	2	3	4	5	6
2	4	6	8	10	12
3	6	9	12	15	18
4	8	12	16	20	24
5	10	15	20	25	30
6	12	18	24	30	36
7	14	21	28	35	42
8	16	24	32	40	48
9	18	27	36	45	54
10	20	30	40	50	60

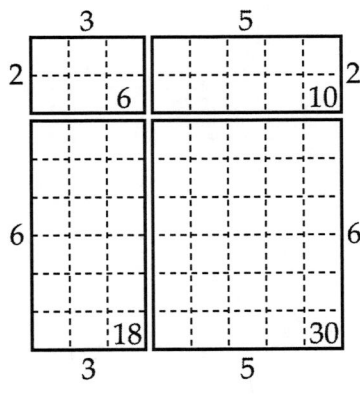

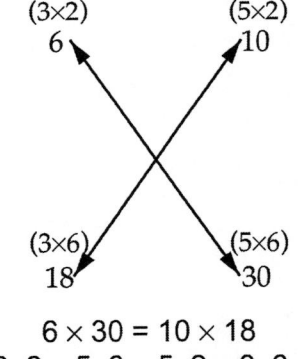

$6 \times 30 = 10 \times 18$
$3 \times 2 \times 5 \times 6 = 5 \times 2 \times 3 \times 6$

$$6 \quad \times \quad 30 \quad =$$
$$(3 \times 2) \times (5 \times 6) = 180$$

and
$$10 \quad \times \quad 18 \quad =$$
$$(5 \times 2) \times (3 \times 6) = 180.$$

Notice that the factors, 2, 3, 5, and 6, that make 180 are the same for 6×30 and 10×18, but in a different order.

You can see it in the figure on the right, where 10 × 18 is the same as six 30s.

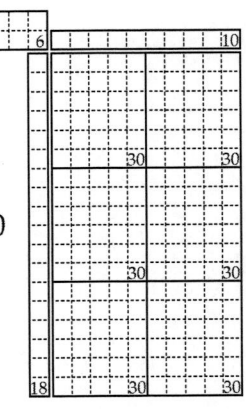

$6 \times 30 = 10 \times 18$

Worksheet 1. Look at the multiplication table on the first worksheet. Try some cross multiplying.

Then do problems 1-3 before reading farther.

Using cross multiplying. You can use cross multiplying to make mental multiplications easier for certain numbers. For example, to find 8×35, first find the 8 and the 35 on the table as shown below in the left figure. Next find the other two squares to complete the rectangle and ratio. See the second figure below, where the 14 and 20 are outlined. Apply cross multiplying: $8 \times 35 = 14 \times 20$; $14 \times 20 = 280$ is easier to calculate than 8×35.

1	2	3	4	5	6	7
2	4	6	8	10	12	14
3	6	9	12	15	18	21
4	8	12	16	20	24	28
5	10	15	20	25	30	35
6	12	18	24	30	36	42

1	2	3	4	5	6	7
2	4	6	8	10	12	14
3	6	9	12	15	18	21
4	8	12	16	20	24	28
5	10	15	20	25	30	35
6	12	18	24	30	36	42

1	2	3	4	5	6	7
2	4	6	8	10	12	14
3	6	9	12	15	18	21
4	8	12	16	20	24	28
5	10	15	20	25	30	35
6	12	18	24	30	36	42

Notice that you could have used a different 8 with the 35. Then, $8 \times 35 = 28 \times 10$, which is easier yet.

Complete the worksheet.

Solving proportions. To solve a proportion without a multiplication table, cross multiply and divide as needed. For example, to solve

$$\frac{18}{24} = \frac{m}{20},$$

cross multiply $\qquad 24 \times m = 18 \times 20$

and divide by 24 $\qquad m = 15$.

Sometimes it is easier to simplify the fractions first as follows.

$$\frac{\overset{3}{\cancel{18}}}{\underset{4}{\cancel{24}}} = \frac{m}{20},$$

cross multiply $\qquad 4 \times m = 3 \times 20$

and divide by 24 $\qquad m = 15$.

Worksheet 2. Problems 5-11 are ordinary proportion problems. Problem 13 has similar triangles, but one triangle is rotated. Be certain to use corresponding sides for your proportion. In Problem 14 you need to realize that this is a special triangle.

The figure for Problem 15 might seem complicated, but the solution is a simple proportion.

Lesson 117

Measuring Heights

GOAL 1. To find the height of a pole, building, or tree with similar triangles

MATERIALS Worksheet 117
Small mirror
A sunny day
Meterstick (or other measuring tool)

ACTIVITIES ***Finding heights.*** In this lesson you will find the height of a pole or tree or a building using three different methods. They all use similar triangles. The first method uses a mirror. The other two use shadows from the sun, which means you will need a sunny day.

Find a height with a mirror. In Lesson 93 you learned that the angle of incidence equals the angle of reflection. In the figure on the right, Morgan is looking into a mirror set on level ground. Morgan walked until the top of the object was visible in the mirror.

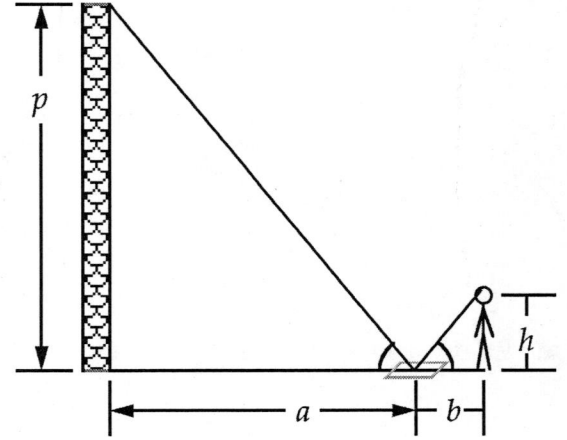

See the similar triangles?

$$\frac{p}{a} = \frac{h}{b}$$

Measure *a*, *b*, and *h*, which is your height from your feet to your eyes.

Find a height with a shadow #1. Find your shadow (or the meterstick's) and the object's. The similar triangles are not hard to find:

$$\frac{p}{s} = \frac{h}{t}$$

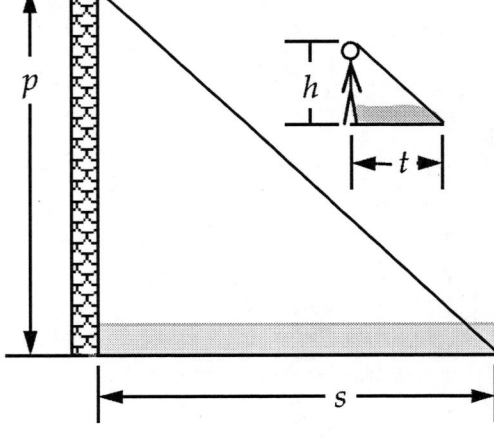

You need to measure both shadows and your true height (or the meterstick's).

> **Whenever your shadow in the sun is less than your height, you need sun protection.**

Find a height with a shadow #2. This time you walk until your shadow meets with the end of the object's shadow. One of the similar triangles is inside the other:

$$\frac{p}{s+t} = \frac{h}{t}$$

Measure *s*, *t*, and *h*.

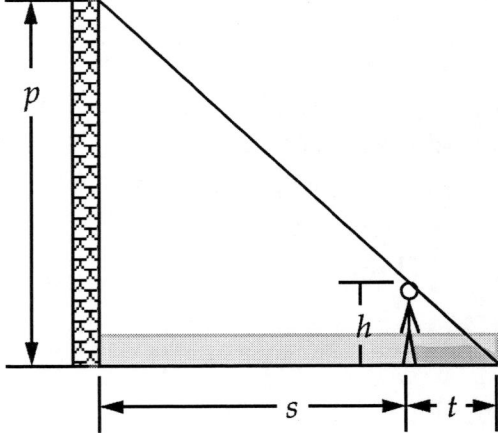

Worksheet. Use the three methods to find the height of an object. Compare results.

Lesson 118

Golden Ratio

GOALS
1. To construct a *golden rectangle*
2. To learn the terms *golden ratio, phi, φ*
3. To calculate the golden ratio

MATERIALS
Worksheet 118
Drawing board, T-square, 45 triangle
mmArc compass, ruler
Calculator

ACTIVITIES

Pleasing rectangles. Which of the four rectangles below do you think are the most eye pleasing? Which ones do you like best?

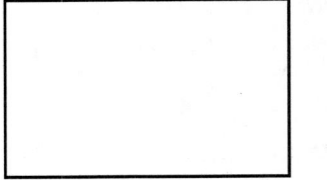 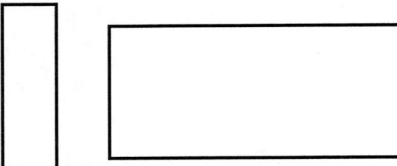 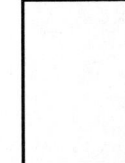

Many people from Western cultures think the rectangles on the ends have the most pleasing proportions. They are *golden rectangles.* Their sides have a special ratio known as the *golden ratio,* or the *golden proportion,* or the *golden section.*

Golden ratio. In the figure below, a line is divided into two parts with the longer part drawn heavier. If a line is divided into two parts with the following proportion

$$\frac{\text{whole}}{\text{longer part}} = \frac{\text{longer part}}{\text{shorter part}},$$

it is called the *golden ratio.*

> To calculate 2 ÷ (√5 - 1), first calculate what's inside ( ) and store in memory. Then do 2 ÷ MR.

Worksheet. Follow the instructions given in Problem 1 and draw a *golden rectangle.* In Problems 2-4 you are to measure to find the golden ratios in the rectangle and the line.

For Problem 5-6, you will be calculating the golden ratio. Complete these before reading any farther.

Phi, φ. Mathematicians have given the golden ratio a Greek symbol, phi, φ(FIGH). See the chart for a comparison with another ratio, π.

> **A riddle:**
> Why does phi + V = 5?
> Because sound of phi + sound of V = sound of 5.

Name	Symbol	Meaning	Approximation
Pi	π	Circumference:diameter ⊖	3.141593
Phi	ϕ	Whole:longer::longer:shorter	1.618034

Problems 7-8. Here you are to verify the amazing ϕ equations:

$$\phi + 1 = \phi^2$$

$$\frac{1}{\phi} = \phi - 1$$

Lesson 119

More Golden Goodies

GOALS
1. To construct a *golden spiral*
2. To find the *golden triangles* in a pentagon
3. To calculate the golden ratio

MATERIALS
Worksheets 119-1, 119-2
Drawing board, T-square, 45 triangle
mmArc compass, ruler
Safe-T Compass®, if available
Calculator

ACTIVITIES
Golden spiral. A golden rectangle can be divided into a square and another golden rectangle. You saw this in the previous lesson. This process can be continued indefinitely, that is, forever.

The left figure below shows a golden rectangle divided into six more golden rectangles. If you draw an arc that is a quarter of a circle in each square, you will have the *golden spiral.* See the right figure below; the black dots show the circles' centers.

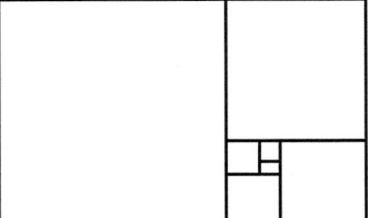 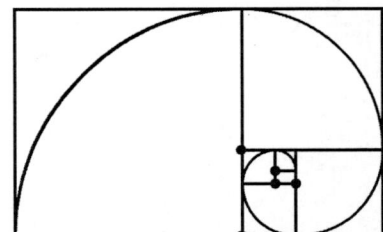

The golden spiral is found in nature in the shell of the chambered nautilus. Astronomers also see it in galaxies. When it is found in nature, it is only an approximation.

Worksheet 1. On the first worksheet you are to draw the golden spiral. The steps are given on the worksheet. Work carefully because in the next lesson, you will be drawing a similar spiral to be compared. You cannot draw the smallest two arcs with the mmArc compass. Use the Safe-T Compass® if you have one.

Worksheet 2. You are to compare the sides of consecutive (neighboring) squares in Problems 2-7. Do Problems 1-7 now.

Golden triangles. A *golden triangle* is first of all an isosceles triangle. And the ratio of the longest side to the shortest side equals ϕ. Two examples are shown in the figures on the right.

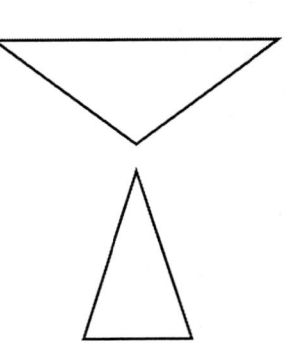

A pentagon with diagonals has golden triangles.

Worksheet 2. In Problems 8-11, you are to determine whether or not certain triangles are golden. The lengths are given so you can calculate ratios to three decimal places. Complete the worksheet.

This fossil and shell are in a museum in England.

Lesson 120

Fibonacci Sequence

GOALS
1. To learn the term *sequence*
2. To learn about the *Fibonacci sequence*
3. To solve some sequence problems

MATERIALS Worksheets 120-1, 120-2, 120-3

ACTIVITIES *Fibonacci.* Fibonacci (fee-buh-NOT-chee), 1170-1250, was born in Pisa (PEE-za), Italy, the city with the Leaning Tower. He was educated in North Africa (now Algeria). Fibonacci learned other mathematics from talking with merchants while traveling around the Mediterranean region.

Fibonacci became fascinated with the Hindu-Arabic numerals, with digits 1-9 and a 0. At that time Europe used Roman numerals, which had no 0. His book *Liber Abaci* ("Book of Calculation") introduced Europe to the numbers we use today. It also showed methods of calculating with paper and pencil without an abacus. Conversion to the new method took time. In 1299 the merchants in Florence required using Roman numerals.

He also introduced the fraction bar—the line separating the numerator and denominator. Before that fractions were written as $\frac{1}{2}$.

Fibonacci sequence. A *sequence* is a set of quantities in some type of order. The answer to one of Fibonacci's math problems results in the Fibonacci sequence. It starts as 1, 1, 2, 3, 5. Think about what comes next before reading further.

If you think 8 comes next, you're right. Each number in the sequence is the sum of the previous two numbers. Thus, $1 + 1 = \underline{2}$, $1 + 2 = \underline{3}$, $2 + 3 = \underline{5}$, and so on.

Problem 1. For Problem 1, you are to practice your addition skills and calculate the Fibonacci sequence to 26 terms. Incidentally, if you learned about check numbers, or casting out nines as they're sometimes called, use them to check your work. Fibonacci learned about them and explained them in *Liber Abaci*.

> If you ever played the "Chain" games, you will recognize the ones column as a chain.

Problem 2. This is the brick wall problem. The sides of the brick have a ratio of 2:1. See the figure at the right. You need to make your wall two units high, but with various widths.

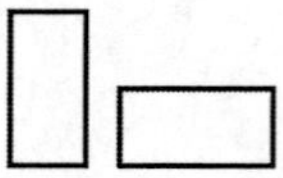

If your wall is 1 unit wide, there is only one way to make it. If it is 2 units wide, there are two ways to make it. See the middle figure below. If it is 3 units wide, there are three ways to make. See the right figure.

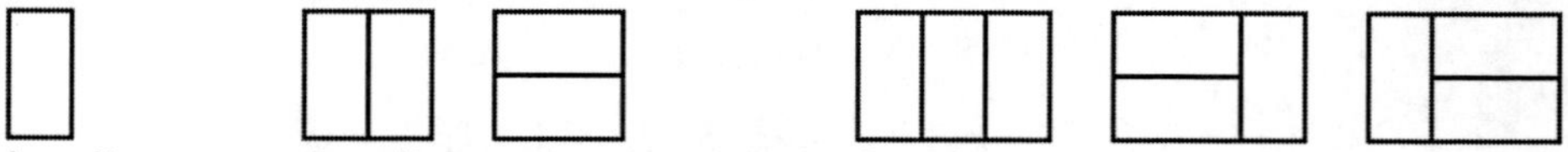

1 unit wide   2 units wide   3 units wide

Continue the process for 4 and 5 units wide. Draw your arrangements freehand and record the number of arrangements. Do Worksheet 1 before reading any further. Think about your solutions.

Discussing Worksheet 1. Notice how quickly the numbers become large in the Fibonacci sequence. To be sure you've added correctly, the last number in the sequence is 121,393.

Discuss with a partner the number of arrangements you found in Problem 2 and why.

Notice you can make the arrangements for 4 units by copying the 3s arrangement plus a vertical brick and copying the 2s plus two horizontal bricks. Likewise, the arrangements for 5 are the 4s plus a vertical brick and the 3s plus horizontal bricks. Adding the last two are, of course, what the Fibonacci sequence is all about. (If you need help in understanding this, look carefully at the solutions.)

Problem 3. For the next problem, you have two sizes of colored rods, the 1s and the 2s. See the figure on the right. You are to make various lengths using these rods. For example, the three arrangements for a length of 3 are shown below. You can do the worksheet now.

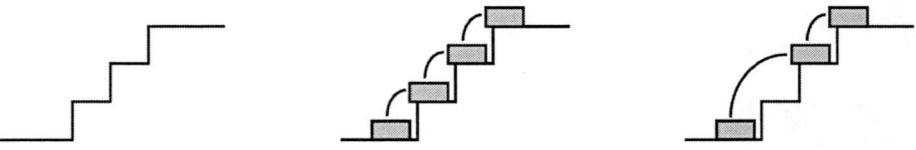

With three 1s   With a 2 and a 1   With a 1 and a 2

Problems 4-5. For Problem 4, you are climbing stairs. You can climb either one at a time or two at a time. The left figure below shows a set of three stairs. The middle figure shows climbing the stairs, one step at a time. The rectangles represent a foot (or shoe); the arcs represent movement.

The right figure shows climbing the stairs by a combination—first two steps at a time and then one at a time. You are to draw all combinations for 3, 4, and 5 stairs.

For Problem 5, the instructions explain the bee problem. The bee can enter a new cell only if the number is higher. The term 123 means "cell 1, then cell 2, then cell 3." Do the worksheet.

New problem. Make up your own Fibonacci problem. If you think of a good one, let me know at joancotter@RightStartMath.com. See some problems at http://www.rightstartgeometry.com.

Lesson 121

Fibonacci Numbers and Phi

GOALS
1. To draw the *Fibonacci spiral*
2. To discover the relationship between Fibonacci numbers and phi
3. To discover relationships within the Fibonacci numbers

MATERIALS
Worksheets 121-1, 121-2
Drawing board, T-square, 45 triangle
mmArc compass, calculator

ACTIVITIES

> *Fibonacci spirals are found in the seed heads of dandelions, daisies, and sunflowers.*

Fibonacci spiral. To draw the Fibonacci spiral, first construct squares whose sides are Fibonacci numbers. Next draw the spiral with arcs, which makes the Fibonacci spiral. It is shown on the right.

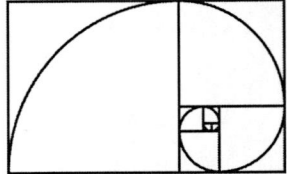

Worksheet 1. Problem 1 guides you in constructing the Fibonacci spiral. Work carefully so you can compare the golden spiral you drew a few lessons ago. Do it now.

Comparing the two spirals. Hold your two spirals together toward a light source. You should notice that the greatest differences occur at the smaller squares. Obviously, there is a connection between the two spirals.

By combining your spirals, you can make all kinds of interesting patterns, as shown at the left and on the previous page.

Problem 2. In 1753, the Scottish mathematician, Robert Simson, discovered a relationship between consecutive Fibonacci numbers and the golden ratio. In Problem 2 you will compare the ratio of consecutive Fibonacci numbers. Either figure out the Fibonacci numbers or use them from the previous lesson. To get the decimal ratios, divide and round to five places.

In the third column, you are comparing how much the ratios differ from the golden ratio. A quick way to find these differences is to use memory on your calculator. First put .61803 into memory. Then to subtract 1 from 1.61803, do the following:

Lamppost in Monte-Carlo, Monaco.

> Press 1.61803 then $\boxed{-}$ $\boxed{\text{MR}}$ $\boxed{=}$. The answer –0.61803 appears. (Ignore the negative sign.)

For the next answer,

> Press 2 then $\boxed{-}$ $\boxed{\text{MR}}$ $\boxed{=}$. The answer 0.38197 appears.

Problems 4-5. In the 1800s, Edouard Lucas made an interesting observation regarding adding consecutive Fibonacci numbers. In Problem 4, n refers to the position in the Fibonacci sequence and f_n refers to the actual number. Thus, when n is 3, f_n is 2 and when n is 6, f_n is 8 and f_{n+1} is 13 and f_{n+2} is 21. You can do Problems 4-5 now.

Problems 6-7. In Lesson 28 you learned that if n is a number, then

$$n \times (n + 2) = (n + 1)^2 - 1.$$

For example,
$$4 \times 6 = 5^2 - 1 = 24.$$

Church door hinge in Strasburg, North Dakota.

In Problem 6 you will discover whether a similar relationship exists with the Fibonacci numbers. Complete the worksheet.

Lesson 122

Golden Ratios and Other Ratios Around Us

GOALS 1. To find the golden ratio in everyday objects
2. To find other common ratios

MATERIALS Worksheet 122
30-60 triangle, 45 triangle
Ruler
Calculator

ACTIVITIES ***History of the golden ratio.*** The mathematics surrounding the golden ratio is true. Euclid wrote about it 2300 years ago in his book *Elements*. Many people believe that the golden ratio has a long history in architecture and art. Unfortunately, most of it cannot be proven, including the story that the designer of the Gizeh pyramid used it.

Leonardo Da Vinci, 1452-1519, is said to have found golden rectangles in the human body and painted with them in mind. There is no direct evidence that this is so. On the other hand, think of all the lines and ratios in a person's face. With so many possible ratios, some are bound to be near the golden ratio.

In fact this special ratio was not even called *golden* until 1835 when Martin Ohm gave it that title.

Problem 1. For Problem 1 you are to measure and calculate the ratios of some common items. Then you decide if it matches the golden ratio.

1618 0339 8874 9894

For the first four problems, you may measure in either inches or centimeters. The answers for the ratios in fractions are given in inches because paper sizes were standardized in inches. Will it make a difference in your ratio decimal answer? If you're not sure, try it both ways.

Problems 2-3. Here you are to measure your drawing triangles and find certain ratios. These are important results that often used in other branches of mathematics. When you use the Pythagorean theorem for the calculated ratio in Problem 3, remember the hypotenuse is across from the right angle.

 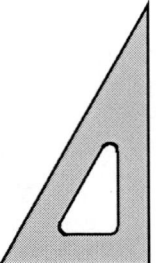

Problems 4-6. No calculations are required. Use either your previous work on the page or what you already know.

Lesson 123

Napoleon's Theorem

GOALS
1. To learn about Napoleon's theorem
2. To construct equilateral triangles with a compass
3. To review constructing medians, altitudes, and angle bisectors
4. To learn the term *generalize*

MATERIALS
Worksheets 123-1, 123-2
T-square, 30-60 triangle
Drawing board, optional
mmArc compass
Goniometer

ACTIVITIES
History of Napoleon's theorem. Napoleon Bonaparte (1769-1821) named himself emperor of France in 1804. Napoleon conquered much of Europe before he was defeated at Waterloo, Belgium, in 1815.

Napoleon was an amateur mathematician. Even through this theorem bears his name, historians are not certain he discovered it.

Napoleon's theorem. Part of Napoleon's theorem is as follows: If equilateral triangles are constructed on the outside of the three sides of any given triangle, then their centroids are the vertices of . . . . You are to finish the theorem after doing Problem 1.

Constructing an equilateral triangle. You already know how to construct an equilateral triangle with a T-square and 30-60 triangle. Now you will learn how to do it with a compass. Follow the steps to draw an equilateral triangle on the line.

1. To find the radius, set the rotator center on one end of the line segment. See left figure. Next move the radius arm to find a hole to match the other end of the line. Remember the hole number.

2. Draw an arc as shown in the second figure with the same hole.

3. Align the rotator center over the other end of the line as shown in the third figure. Use the same hole and draw an arc to intersect the first arc.

4. Connect the ends of the line segment to the intersection of the arcs. See the last figure.

> **See Lesson 52 if you need a reminder about centroids.**

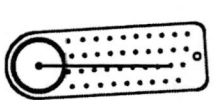
Align the mmArc compass.

Draw an arc.

Draw another arc.

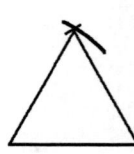
Draw lines.

Worksheet 1. Follow the instructions for Problem 1. Notice the result and complete the theorem for Problem 2.

For Question 3, take a careful look at the three external triangles. Do some measuring to verify your guess.

Problem 4 asks you find centroids. If you've forgotten what a centroid is, refer to your glossary or Lesson 52. After some investigating, answer Question 5.

Worksheet 2. For Problem 6, use your T-square and a triangle to construct the altitudes. It is your choice which altitudes to draw in each triangle.

Problem 8 needs angle bisectors. If necessary, refer to Lesson 71.

Do the worksheets before reading farther.

Generalizing. To *generalize* in the mathematical sense means to extend an idea. Napoleon's theorem only referred to constructing equilateral triangles around a given triangle. Mathematicians were naturally curious to see what would happen if they tried triangles that were not equilateral.

So they tried similar triangles. They found they worked. (They have to be constructed so a different side touches the given triangle –rotated about the given triangle.) It didn't produce an equilateral triangle, but a similar triangle. That was a generalization from equilateral triangles to similar triangles.

Next mathematicians thought about the centroids. Remember that in an equilateral triangle, the intersection of the medians, the altitudes, and the angle bisectors are the same point. So the question was would the other points work. You know the answer. That was another generalization, this time from centroid to other triangle points.

A tessellation based on Napoleon's theorem.

Lesson 124

Pick's Theorem

GOALS
1. To determine Pick's theorem
2. To apply Pick's theorem for finding areas
3. To review finding areas for trapezoids and parallelograms

MATERIALS
Worksheets 124-1, 124-2
Drawing board, T-square, 45 triangle
Geoboard, if available

ACTIVITIES
Pick's theorem. Mathematician Georg (GAY-ork) Pick was born in 1859 in Vienna, Austria, and died in the Theresienstadt concentration camp around 1942 or 1943. Pick's theorem was published in 1899, but not generally known until 1969.

Pick's theorem finds the area of a polygon on a grid such as a geoboard. It involves counting the boundary points and interior points. A boundary point is a point on a side of the polygon. An interior point is a point inside the polygon. See the figure below.

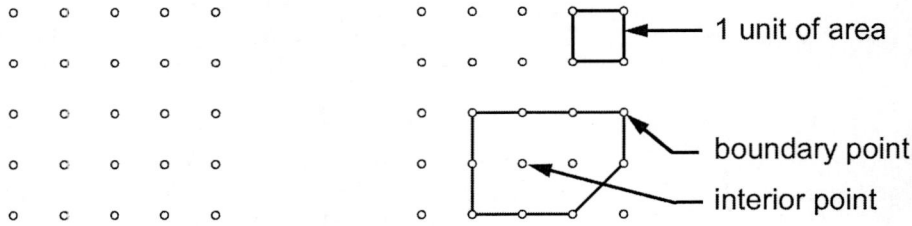

Square grid, also called a lattice.

Worksheet 1. On the first worksheet, you are to figure out Pick's formula for finding area. First you will consider cases without interior points. You will do this in Problems 1-4. You can find the areas by looking carefully at the figures. Answer Questions 2 and 3 thoughtfully because that is the thinking you need to find the formula.

In Problems 5-7 you will generalize your formula to include interior points. The result is known as Pick's theorem.

Geoboard. If you have a geoboard, make some figures and find their areas. First make some simple shapes that you can easily find the areas. Then use Pick's theorem to see if they agree. Next make some complicated figures and use Pick's theorem to find the areas.

Worksheet 2. You might need to look up the formula for the area of a trapezoid for Problem 7. Your answers for finding the area using the traditional formula and Pick's theorem better agree.

For Problem 8, you will need to draw eight different–not congruent– triangles. Each vertex must be on a lattice point. Find their areas mentally by looking at the number of squares, or by Pick's theorem, or by using the usual formula.

Problem 9 is similar to Problem 7 in that you are to use two methods for finding the areas. Before you start, do you know which one has the greatest area.

Pick's Theorem With the Stomachion

Lesson 125

GOAL 1. To learn about Archimedes and the Stomachion

MATERIALS Worksheets 125-1, 125-2
Drawing board, T-square, 45 triangle

ACTIVITIES **Archimedes.** Archimedes (ar-ki-MEE-dees) is considered the greatest scientist and mathematician of the ancient world. He also is considered one of three greatest mathematicians of all time.

He was born in Syracuse, Sicily, in 287 B.C. At that time Syracuse was a 500-year-old Greek city-state. A Roman soldier who didn't know who he was killed him during an attack on the city in 212 B.C.

An ancient manuscript. Copies of Archimedes' work exist today. One copy, made in the 900s, is especially interesting because of its recent history. In the 1200s, monks in Constantinople (now Istanbul, Turkey) recycled the manuscript by scraping off much of the original math and writing Christian prayers over it.

In 1907 a Danish historian, Johan Ludvig Heiberg, discovered the 174-page manuscript in a monastery library in Istanbul. Heiberg photographed each page before the manuscript went missing.

It reappeared at an auction in New York in 1998. An anonymous billionaire bought it for $2 million and loaned to a museum in Baltimore, Maryland. Restorers and scholars, using high-tech digital imaging, are able to decipher much of it.

> *The PBS TV program, NOVA, broadcast a show about this manuscript, "Infinite Secrets," on September 30, 2003.*

Stomachion. On the ancient manuscript was a section about an ancient puzzle, called the Stomachion (sto-MOCK-yon). This 14-piece puzzle will form squares or many other shapes, such as an elephant. See figures at the right. It is a super tangram.

The modern world didn't know that Archimedes had worked on the Stomachion. At first mathematicians wondered why Archimedes was interested in a puzzle. Now we know he wanted to figure out how many different ways the pieces could form a square. Finding the answer was not easy.

On October 31, 2003, Bill Cutler modified a computer program and found the number of squares to be 536. Another website gives various dancers made from all 14 pieces. You can see both at http://www.rightstartgeometry.com.

Stomachion and Pick's theorem. In 2003-2004 Athina Markopoulou, an electrical engineer, working at Stanford University in California discovered that all the square solutions fit on a 12 × 12 lattice. She is now working to keep the Internet operating smoothly.

Worksheet 1. Before tackling the Stomachion, you will work with a tangram and squares on a lattice. To calculate the area for *n* in Problems 4-5, compare the values for sides 1-4 with corresponding areas.

Worksheet 2. For Problem 19 there are five trapezoids; name three.

Athina Markopoulou

Lesson 126

Pick's Theorem and Pythagorean Theorem

GOALS 1. To learn to draw squares on oblique lines on a lattice
2. To verify the Pythagorean theorem with Pick's theorem

MATERIALS Worksheets 126-1, 126-2
Drawing board, T-square, 45 triangle
Calculator, ruler

ACTIVITIES ***Drawing a square given a line.*** The problem is to make a square with $\overline{AB}$ one of the sides. Start by drawing a line at point A. How do you know which lattice point to choose for point C? Look at the left figure below.

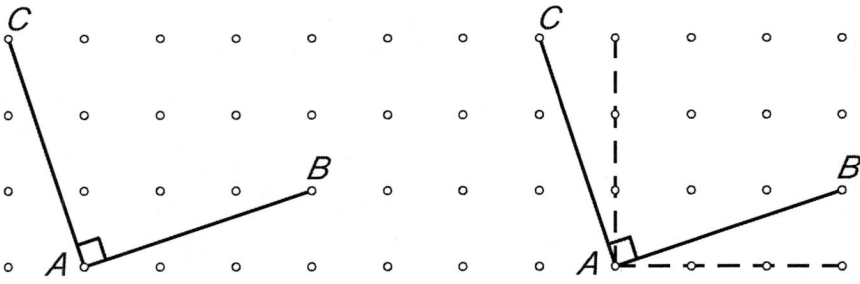

If $\overline{AB}$ were horizontal like the dotted line in the right figure above, the problem would be easy. You'd simply rotate the line 90° and draw the vertical dotted line.

> **If you have forgotten about the Pythagorean Theorem, see Lesson 57.**

Compared to point A, point B is 3 spaces to the right and 1 space up. Using coordinates we write it as (3, 1). Notice that point C is 1 space left and 3 spaces up, or (–1, 3). Same numbers, different order! All this sounds more complicated than it really is. But you will need this to make the necessary squares for the worksheets. See an example in the figure below.

Worksheets. For Problem 1, draw squares on the three sides of the triangle. You could check with your drawing tools to be sure your squares are square. Next find the areas of the squares on the legs. Record them in the chart in the columns "Area Leg1" and "Area Leg2." Add them and record the area in "Area Legs."

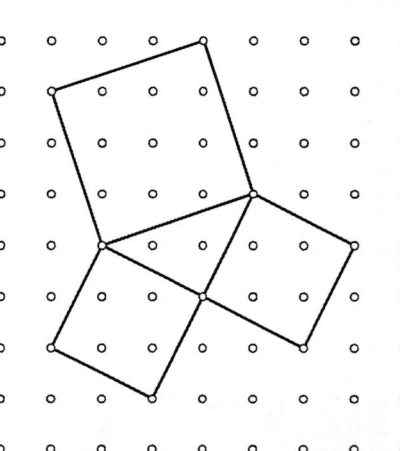

Next find the area of the square on the hypotenuse and record it under "Area Hyp."

The fourth column is the calculated length of the hypotenuse. If you know the area of a square, how do you find the length of a side? The answer is at the bottom of the page.

For the last column, measure the hypotenuse in centimeters. How close was your calculation to your measurement?

Repeat for the remaining six triangles. [Ans. find the square root]

Lesson 127

Estimating Area With Pick's Theorem

GOAL 1. To estimate areas with Pick's theorem

MATERIALS Worksheet 127
Drawing board, T-square, either triangle
Ruler

ACTIVITIES ***Estimating with Pick's theorem.*** Below in the left figure is a circle. The area using the standard formula is 12.57 unit^2. You can use Pick's theorem to approximate the circle's area. First you need a set of lattice points. See the second figure. A grid is much easier to draw than a parade of dots. The lattice points are, of course, the intersections on the grid.

Next use the lattice points and draw a polygon as close to the circle as possible. See the right figure.

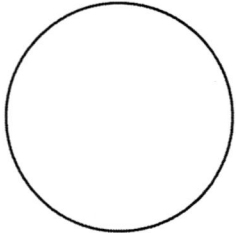

$A = \pi r^2$

$A = (2)^2$

$A = 12.57 \text{ unit}^2$

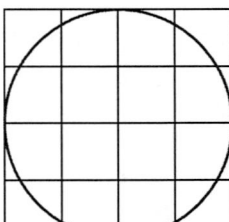

$A = \dfrac{B}{2} - 1 + I$

$A = \dfrac{11}{2} - 1 + 8$

$A = 12.5 \text{ unit}^2$

The area calculation, 12.5 unit^2, with Pick's theorem is very close to the actual area.

Problem 1. You are given an ellipse on a grid. Draw a polygon that is as close as possible to the ellipse. Each vertex of the polygon must be on a lattice point. Try to draw your polygon with a balance between the amount of area outside the ellipse and the amount inside.

Then calculate the area using Pick's theorem. The actual area is 125.7 unit^2. Do it now and see how close you come.

Problem 2. In Problem 2, you are to find the area of the given Maltese cross. Use the unit square shown, which is placed in the center of the cross. Use your ruler and mark off points for the remaining lines. Then use your drawing tools to complete the grid. Finally use Pick's theorem to find the area.

Problem 3. For this problem you decide where to place your grids.

*Problem 4. Find a map of your state or province. Use the same procedure discussed in Problem 3 and find its area. Compare it to the actual area.

The area of wildfires have been calculated by using this method.

Lesson 128

Distance Formula

GOALS 1. To learn the distance formula using the Pythagorean theorem
2. To find distances in a coordinate system

MATERIALS Worksheet 128
Drawing board, T-square, 45 triangle, optional
Calculator

ACTIVITIES ***Distance between two points.*** A common problem is to find the distance between two points, *AB*. Sometimes the problem is on a lattice, or grid, such as city streets. See the left figure below.

You could solve the problem by constructing a square on *AB* and then using Pick's theorem to find the area and take the square root. That is not very practical for large numbers; there is an easier way. You know that squares drawn on the legs equal the square on the hypotenuse. How about making the legs horizontal and vertical? See the second figure below.

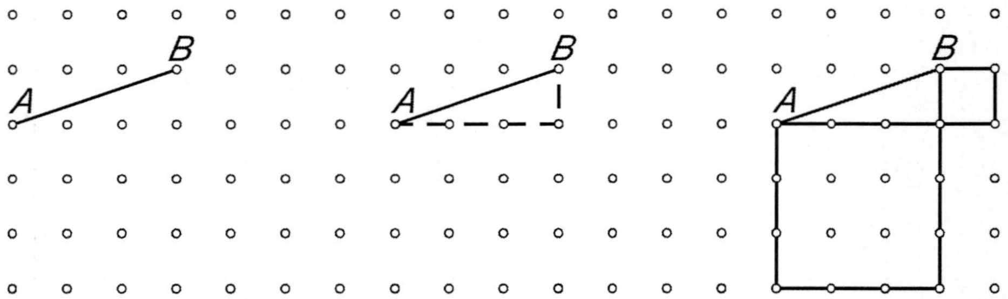

Distance formula. If *d* is the distance, or length, of a line segment and *x* and *y* are the horizontal and vertical lengths, the formula is:

$$d^2 = x^2 + y^2$$

So *d* is:

$$d = \sqrt{x^2 + y^2}$$
$$= \sqrt{3^2 + 1^2}$$
$$= \sqrt{10} \approx 3.162$$

The exact answer is the square root of 10, which is approximately 3.162 to three decimal places.

Distance formula with coordinates. A coordinate system is similar to a lattice. To find the distance between two points, first plot the two points and draw the line between them. Then use the distance formula. Remember that the *x* is the total length in the *x* direction, and *y* is the total length in the *y* direction. See the figure on the right. The *x* value is 4 and the *y is 2.*

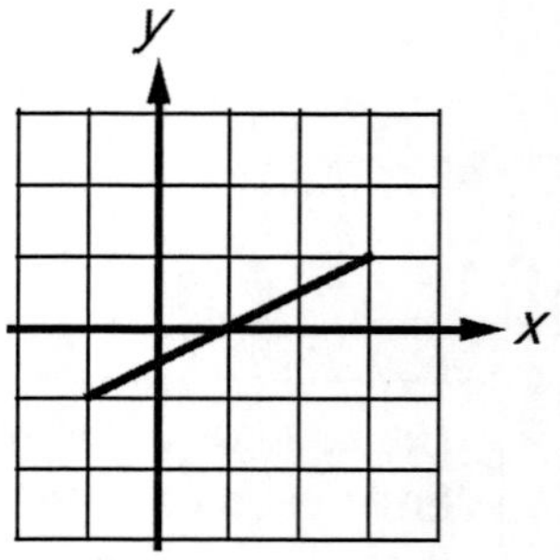

Worksheet. Problems 1-3 are on a lattice. Visualize the squares that could be constructed on the legs of the right triangles. Problems 4-8 use a coordinate system. Solve them the same way.

Lesson 129

Euler Paths

GOALS
1. To learn to identify a *Euler path*
2. To solve some problems using Euler paths

MATERIALS Worksheets 129-1, 129-2

ACTIVITIES ***Euler path.*** An *Euler* (OIL-er) *path* is a continuous path that travels on every line or arc in a figure without repeating. For example, it is obvious that you can trace the square shown on the left. But no matter how hard you try, it is impossible to trace a square with its diagonals. See the two attempts shown on the right.

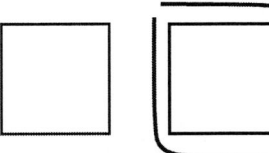 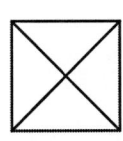  

In the middle figure, one diagonal is skipped. In the right figure, the bottom line is traced twice.

> A Swiss mathematician, Leonhard Euler, 1707-1783, figured out how to solve this problem.

Even vertex and odd vertex. To discuss these paths, you need to know the difference between an even vertex and an odd vertex. An even vertex has an even number of lines or arcs leading to it. An odd vertex has an odd number. The figures below show both types.

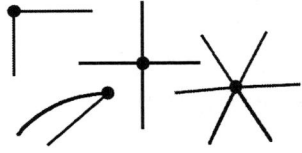 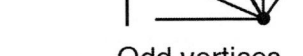

Even vertices   Odd vertices

Worksheet 1. For Problem 1 you are to decide whether or not each figure has an Euler path. If it has, figure out how many different vertices you can start from. Also write down the number of even and odd vertices.

Do the worksheet now before reading further. Discuss your answers with a partner and compare with the solutions.

Problems 2-8. Answer these questions carefully and see if you can discover which figure has an Euler path. Answer them before reading any farther.

Summary. There are three parts to Euler's theorem about paths:

1. Figures with only even vertices have Euler paths. You can start at any vertex to draw a path. You will end where you start.

2. If a figure has more than two odd vertices, it does not have an Euler path.

3. If a figure has two odd vertices, it has Euler paths. One odd vertex is the starting point and the other is the ending point.

Problems 9-11. Problem 9 is not hard now that you know the theory. For Problems 10-11, if a path if possible, draw it and show the starting and ending points.

Lesson 130

Using Ratios to Find Sides of Triangles

GOAL 1. To set up and solve equation using a given ratio

MATERIALS Worksheet 130
Drawing board, either triangle
Calculator, goniometer, ruler

ACTIVITIES ***Problem 1.*** Use the figure below and the ratios given to solve the following problem: If $c = 10$, find a. Solve the problem before reading further.

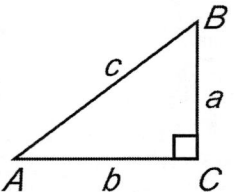

$$\frac{a}{c} = .60 \qquad\qquad \frac{a}{b} = .75$$

Since we are working with a and c, we need the $\frac{a}{c} = .60$ ratio.

Substitute $c = 10$,
$$\frac{a}{10} = .60$$

Either think of multiplying by 10 on both sides
$$a = 10 \times .60$$
$$a = 6$$

or, cross multiply
$$\frac{a}{10} = \frac{.60}{1}$$

Sometimes, textbooks write a 1 under a number to make it look like a fraction. This is unnecessary and not a good habit.

Problem 2. Use the same figure and ratios: $a = 12$, find b. Solve it two ways and compare with a partner. Then continue reading.

Use the $\frac{a}{b}$ ratio
$$\frac{12}{b} = .75$$

Ratios can be "turned," for example,
$$\frac{2}{4} = \frac{3}{6} \rightarrow \frac{2}{3} = \frac{4}{6}$$

Turn the original equation
$$\frac{12}{.75} = b$$
$$b = 16$$

Second method:
$$\frac{12}{b} = .75$$

cross multiply
$$12 = .75 \times b$$

and divide by .75
$$\frac{12}{.75} = b$$
$$b = 16$$

Worksheet. Problems 1-8 are ratio problems. A calculator is necessary for 6 and 7. Be sure your answers make sense. The remaining questions ask you to think about ratios and similar triangles.

Lesson 131

Basic Trigonometry

GOALS
1. To learn the meaning of *trigonometry*
2. To learn the terms *opposite, adjacent, sine, cosine,* and *tangent*
3. To construct trigonometry tables

MATERIALS
Worksheets 131-1, 131-2
Calculator
Goniometer, ruler

ACTIVITIES
A trigonometry problem. In the problem below, you are given *c* and ∠A and asked to find *a*. If you knew the ratio, $\frac{a}{c}$, you could solve the problem as you did on the previous worksheet.

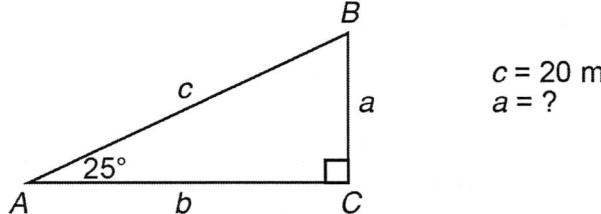

Solving problems using triangle ratios involves a branch of mathematics called *trigonometry* (TRIG-ah-NOM-ah-tree). The name is often shortened to *trig*. The derivation of the word is quite simple. *Tri* means *three, gon* means *angle,* and *metry* means *measure.*

Trig terms. Basic trig uses six terms, one you already know, two are simple, and three will be new. Look at the triangle below; the hypotenuse is opposite the right angle. $\overline{BC}$ is the leg *opposite* ∠A and $\overline{AC}$ is the leg *adjacent* to ∠A.

> *You might like this for remembering the trig ratios: soh-cah-toa. (For example, "soh" is sin, opp, hyp.)*

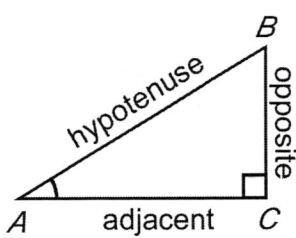

$$\text{sine }(A)=\frac{\text{opposite}}{\text{hypotenuse}}$$
$$\text{cosine }(A)=\frac{\text{adjacent}}{\text{hypotenuse}}$$
$$\text{tangent }(A)=\frac{\text{opposite}}{\text{adjacent}}$$

$$\sin (A)=\frac{\text{opp}}{\text{hyp}}$$
$$\cos (A)=\frac{\text{adj}}{\text{hyp}}$$
$$\tan (A)=\frac{\text{opp}}{\text{adj}}$$

The trig names for the ratios are given in the first rectangle above. They are usually abbreviated as given in the second rectangle. However, the abbreviations are pronounced like the original words, sin (SIGN), cos (KOH-sign), and tan (TAN-jent).

Trigonometry history. The Babylonians, Greeks, Egyptians, Indians, Chinese, and Arabs all contributed to the field of trigonometry. Surveyors and astronomers use it extensively for finding distances.

Worksheets. On the worksheets, you will be making a trig table for angles between 5° and 85°. On Worksheet 1 use the Pythagorean theorem and calculate the ratios to 3 decimal places. On Worksheet 2 measure the sides with a ruler and the angles with a goniometer. Then calculate the ratios and enter them in the table on Worksheet 1.

Lesson 132

Solving Trig Problems

GOAL   1. To solve various problems using trigonometry

MATERIALS   Worksheet 132
Calculator
Drawing board, either triangle

ACTIVITIES   ***Worksheet.*** Go ahead and try to solve the four problems. For the trig ratios, use the solutions for your table from the previous lesson.

If you need help, then refer to the discussions below. Even if you solve them all without help, you might want to read the write-ups afterward.

Problem 1. This is a plain ratio problem. To decide which ratio you need, ask what you know and what you want. You know the adjacent leg and you want the opposite leg. So the trig ratio you need is the tangent.

Note that your answer may not quite agree with the solutions. Trig ratios cannot be calculated very accurately by measuring as you did.

Problem 2. Here you have the hypotenuse and you need both the opposite and adjacent legs. The sine and cosine are what you need.

Once you found a, you could have used the tangent. But it will not give the most accurate answer because you are using a computed value, a, rather than an original value of 36.

Problem 3. At first you might be tempted to use the Pythagorean theorem, but it's not necessary. What do you need to use a trig ratio? You need an angle, which you can get by finding the ratio and then using the trig table.

Now you have the angle and two sides. Choose the trig ratios that include the opposite leg. Your answers probably will not agree, again because your table isn't very accurate.

Problem 4. To find the area of a triangle, you need the width and the height. The width you have. How do you get the height? First draw it and then you'll see another triangle you can solve.

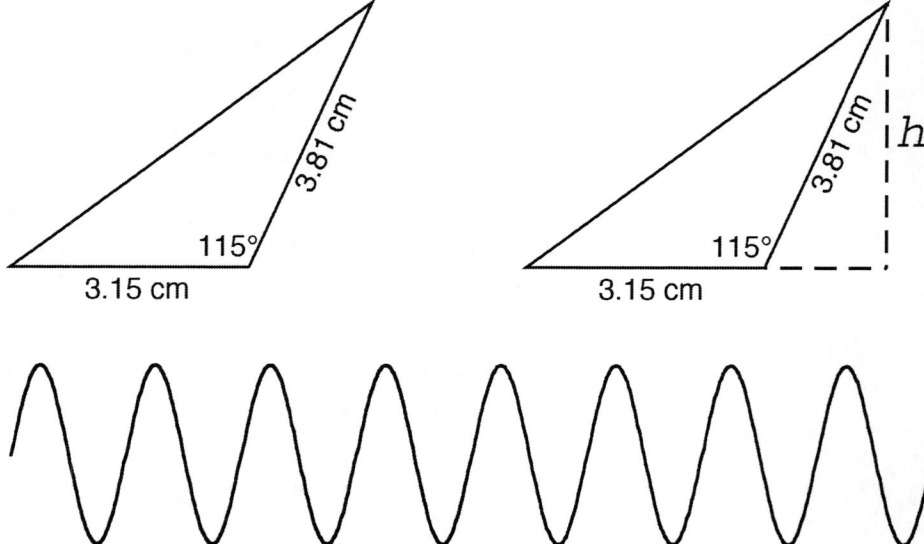

Lesson 133

Comparing Calculators

GOALS
1. To learn some differences between a basic and *scientific calculator*
2. To learn some features of the scientific calculator

MATERIALS
Basic calculator (Casio SL-450L)
Scientific calculator (Casio fx-300MS)
Worksheet 133

ACTIVITIES
Calculators. There are dozens of different types of calculators. A scientific calculator does the basic operations that a simple calculator does, but in a slightly different way. In addition to basic arithmetic, a scientific calculator does trigonometric functions, exponents, means (averages), has () and pi, and many more advanced topics.

The Casio fx-300MS may be used on the SAT and PSAT/NMSQ tests.

 Basic calculator

 Scientific calculator

Before using your new calculator, read the Safety Precautions and the Handling Precautions. Find them on the User's Guide. The best way to learn to use it a new calculator is to experiment with it. Glance at the instructions when you need help with more advanced features.

If your calculator was used for advanced work, press MODE (next to ON) and 1 and =.

Order of operations. Turn on the scientific calculator by pressing the [ON] button in the upper right corner. Perform the following operations in the order it is written on both calculators

$$2 \boxed{+} 3 \boxed{\times} 4 \boxed{=} \underline{\quad}$$

One calculator says 14 and other says 20. The correct answer is 14, according to the order of operations, which says multiply and divide before adding. Scientific calculators and spreadsheets follow the standard order of operations.

How did you like seeing all the numbers you entered?

Rounding. Enter 2 ÷ 3 on both calculators. Notice the scientific calculator has a string of 6s, but ends with a 7. It rounds the answer. The basic calculator has a string of 6s. It does not round, but just quits, which is called truncating.

As another example, try

$$1 \boxed{÷} 3 \boxed{+} 2 \boxed{÷} 3 \boxed{=} \underline{\quad}$$

What should the answer be? The scientific calculator sees it as two fractions, or division, added together, ⅓ + ⅔. Try it. On the other hand, the basic calculator does 1 ÷ 3, adds 2, and then divides all of that by 3. To get the correct answer on the basic calculator, you must use memory. First do 1 ÷ 3 and add to memory. Next do 2 ÷ 3 and add to memory. Try it and you'll find the answer still isn't exactly right; it says .99999999, instead of 1.

Square root. On the basic calculator you found a square root by entering the number and $\boxed{\sqrt{}}$. On this scientific calculator, you first press $\boxed{\sqrt{}}$, then enter the number, and press $\boxed{=}$.

Enter 13 and find its square root on both calculators.

> Basic: 13 $\boxed{\sqrt{}}$ = 3.6055512
> Scientific: $\boxed{\sqrt{}}$ 13 = 3.605551275

Now square your results. On the basic calculator, press $\boxed{\times}$ and $\boxed{=}$. On the scientific calculator, press $\boxed{x^2}$ and $\boxed{=}$. Do they both say 13? Also note that the basic calculator truncates the answer. If it had rounded it, the last digit would have been 3, not 2.

Pi. On your basic calculator, you had to enter several digits to find pi. On a scientific calculator, π is built in to 10 places.

You will find the π symbol in the middle of the bottom row, above the $\boxed{\text{EXP}}$ key. To use any function printed in gold letters, first press the $\boxed{\text{SHIFT}}$ key in the upper left hand corner. Therefore, to access pi, press $\boxed{\text{SHIFT}}$ and $\boxed{\pi}$; press $\boxed{=}$ to see the actual value.

Memory. To place the value on the display into Memory, press $\boxed{\text{SHIFT}}$ $\boxed{\text{STO}}$ $\boxed{\text{M}}$. To multiply by the value in Memory, press the number and then $\boxed{\times}$ $\boxed{\text{RCL}}$ $\boxed{\text{M}}$ $\boxed{=}$. The values in Memory remain even after the power is off.

Worksheet. The 10 problems are actually geometry review problems. Do them on your scientific calculator without writing anything down. Be sure to estimate the answer and check that it makes sense. solve them now. When you have finished or need help, continue reading below.

Problem 1-2. These are plain problems involving π. Don't confuse diameter and radius.

Problem 3. There are two ways to solve this problem. Try both.

Problem 4. Here is where you'll appreciate the scientific calculator.

Problem 5. You really don't need the calculator for this one.

Problem 6. This would be a tough problem without a calculator. After finding the ratio, store it in Memory. Then enter each side of the polygon and press $\boxed{\times}$ $\boxed{\text{RCL}}$ $\boxed{\text{M}}$ $\boxed{=}$.

Problem 7. Remember something special about isosceles triangles.

Problem 8. The easy way to do this problem is to recognize the relationship between the two squares. Mentally fold a corner of the larger square unto the smaller square.

Problem 9. Using parentheses makes this problem easier.

Problem 10. This is a two-step problem, not hard if you remember what you're doing.

> **When the calculator shows, "Syntax Error," press the red button, $\boxed{\text{AC}}$, and try again.**

Lesson 134

Solving Problems With a Scientific Calculator

GOALS 1. To learn to use a scientific calculator to solve trig problems
2. To solve problems using trigonometry

MATERIALS Worksheet 134
Scientific calculator Casio fx-300MS

ACTIVITIES ***Trig functions on a scientific calculator.*** You can find trigonometric functions on a scientific calculator. Find sin (30) by pressing (SIN) 30 (=). The results should be 0.5. If it says –0.988031624, you need to change to degrees. (Angles are often measured in radians for advanced mathematics.) Change it to degrees by pressing (MODE) several times until you see "Deg Rad Gra." Then press 1.

Inverse functions. In the Problem 3 in Lesson 132, you needed to find an angle when you had the cos of the angle. On a calculator the inverse, or opposite, of cos is written cos⁻¹. For example, find the angle whose cos is .5. First activate cos⁻¹ with (SHIFT) (COS⁻¹), then press .5 (=). The answer should be 60. The inverse of sin is sin⁻¹ and the inverse of tan is tan⁻¹.

There are two ways to find the angle when the sin of the angle is a ratio. For example, find the angle when the sin of the angle is $\frac{\sqrt{3}}{2}$. One way is to use the ANS button. First press (√)3 (÷)2 (=). Then press (SHIFT) (SIN⁻¹) (ANS) (=). The answer will be 60.

Or, use parentheses: press (SHIFT) (SIN⁻¹) (() (√)3 (÷)2 ()) (=).

Problems 1-4. The first four problems ask you to solve the four problems from two lessons ago with a scientific calculator.

In the first two problems, your answers will be more accurate because you have the trig values to more places.

In the third problem, put your angle answer (it is not exactly 75°) in memory. Then find the sin (or tan) of the angle and continue with multiplying. You will find the two answers for *a* to be the same.

In Problem 4, after you have found the height, record it as a rounded value. But keep the entire value of the height on your calculator to find the area. Then round your area answer. This way your answer will be more accurate.

Problem 5. To find the angle, you will need to use the inverse tan function. Record your angle answer. Then use the complete angle value to continue your calculations.

Problem 6. In the U.S. and Canada, roofing materials are measured in sq. ft. So this problem uses feet and inches. Roof pitch is measured by the amount of rise per 12" of run. The roof shown is the figure on the right has a pitch of 8:12. Does it make any difference whether the rise and run is measured in mm, cm, in., or in paper clips? The answer is at the bottom of the page.

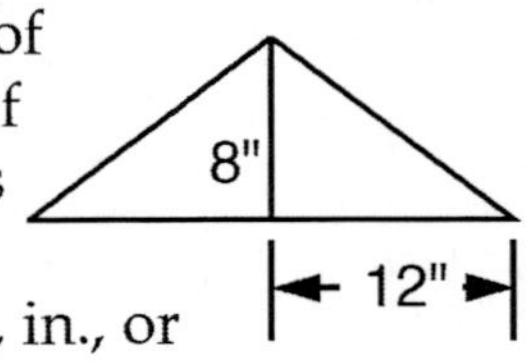

In discussing the pitch of a roof, it is customary to use the amount of rise, or height, in inches for every horizontal 12 inches. [Answer: no]

Lesson 135

Angle of Elevation

GOALS
1. To learn the terms *angle of elevation* and *stride*
2. To build a simple *clinometer*
3. To use the clinometer to measure angles
4. To find remote heights and distances

MATERIALS
Goniometer, optional
Worksheets 135-1, 135-2
Scissors
String or dental floss, small weight, such as a key. (15 cm of floss extending from a small dental floss container makes a good string and weight.)
One or two 2 cm binder clips, or glue, glue stick, tape, or staples
Scientific calculator
Measuring tape

ACTIVITIES
Angle of elevation. In the figure below, a person is holding a goniometer at eye level. The lower bar is horizontal and the upper bar is pointed at the top of the wall. The angle on the goniometer is the *angle of elevation.*

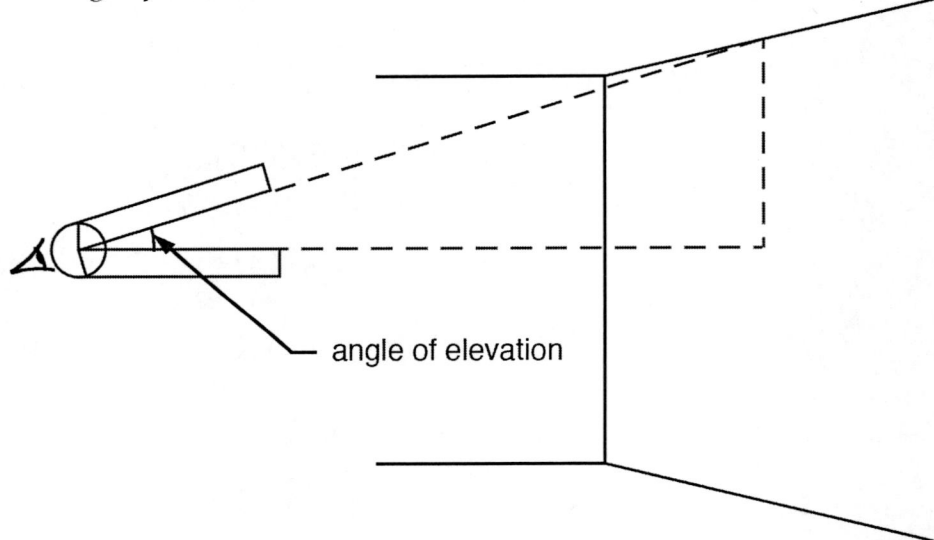

angle of elevation

You can find the height of the wall, h, if you know the angle of elevation and how far you are from the wall. In the figure on the right, you know the angle and d. So, to find h, you use the tangent of the angle.

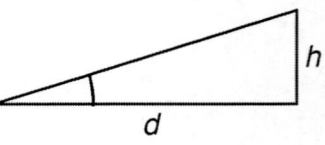

Obviously, to find the wall's height, you also need to add your eye height.

Clinometer. There is a problem that makes a goniometer unsuitable for measuring the angle of elevation. How do you keep the lower bar horizontal?

A *clinometer* (kly-NOM-i-ter) is a tool for measuring such angles. Since they are expensive, you will make a simple one out of paper.

Worksheet 1. The first worksheet has the clinometer and instructions for assembly. Construct it now.

Measuring an angle of elevation. Look through your clinometer at the top of the wall. When you are looking in the viewing tube, line up your object with the bottom edges of the tube.

If you have a partner, ask him or her to read the angle once it stops wiggling. If you are working solo, use your thumb to hold the string until you can read it. You can move the string to either side of your clinometer. But be sure the string's knot is over the hole.

Finding a height. Think about how you'd solve the following problem before reading the solution.

In a certain building, Jamie wants to find the height of the ceiling in feet. The angle of elevation is 14.5° and the distance from the point where Jamie measured to the wall is 154". Jamie's eye height is 58.3".

Jamie drew a sketch and then did the calculations.

$$\tan 14.5 = \frac{h}{154}$$

$$h = 154 \times \tan 14.5$$

$$h = 39.8\,"$$

w is distance from floor. $w = 39.8 + 58.3 = 98.1"$

Since 12" = 1' $w = \dfrac{98.1}{12} = 8.2'$

The value for h is rounded, but the full 10 digits are kept on the calculator for the next calculation, w. Also, the full 10 digits of the value for w are divided by 12.

Measuring with strides. Rather than measuring distances with a meterstick or tape, you can measure with your stride. A stride, mentioned in Lesson 113, is the distance your foot travels when walking until it touches the ground again. See the figure below.

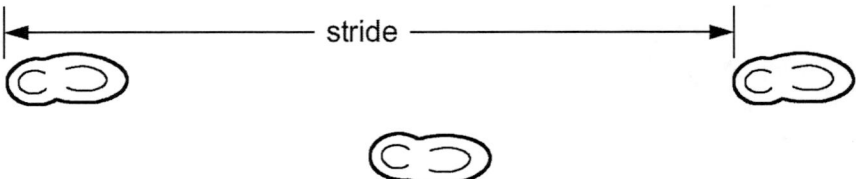

Then to measure a distance, you count the number of strides, multiply by your stride length, and add your partial stride.

Worksheet 2. Find a tall structure outside, for example, a flagpole, tree, or a building. Mark your standing point. Measure your distance from your object, using either a measuring device or strides. Measure the angle of elevation. Do it from two places.

Problem 3. Find a tall object to measure. Measure its height from two different angles. See how close your answers are. You may use strides if you wish. Complete the worksheet.

The top of the Concord Point Lighthouse at Havre de Grace, MD. Notice it has nine sides. John Donohoo built it in 1827.

Lesson 136

More Angle Problems

GOALS
1. To find the angle of the sun
2. To learn the term *angle of depression*
3. To solve more problems using trigonometry

MATERIALS
Clinometer
Worksheet 136
Scientific calculator

ACTIVITIES

A clock in the Orsay Museum in Paris, France.

Supporting beams for the Eiffel Tower in Paris, France.

An altimeter, which measures altitude.

A Ferris wheel in Yate, Bristol, England.

The angle of the sun. You can use your clinometer to measure the angle of the sun. Do NOT look at the sun through the viewing tube.

Instead, point the "string" end of the clinometer towards the sun. Place your other hand perpendicular to the other end of the viewing tube about 1 cm away. Adjust the clinometer until you see the sun shining through the tube. The figure at the right shows what you should see on your hand. It helps to line up the binder clip handles. Then read the angle of the sun.

Using the angle of the sun. The sun's angle automatically gives you the angle of elevation without any correction for eye height. To find the height of the tree, measure the sun's angle and the tree's shadow. Be sure to take both measurements as close as possible to the same time.

Angle of depression. You can use your clinometer to measure angles that are lower than you. See the figure below where the person at the window is measuring an angle to a point in the street. Notice that the angle of depression is the same as the angle of elevation. As you know from Lesson 47, the angles are alternate interior angles.

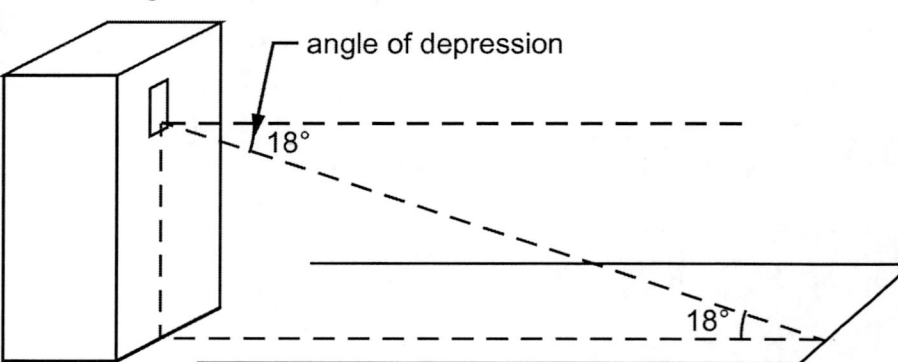
angle of depression

18°

18°

To measure angles of depression, place the "string" end near your eye. See the figure at the right.

Worksheet. For Problem 1, measure the angle of depression for an object lower than you. Draw a sketch and solve your problem.

Also draw pictures for Problems 2-4. Then solve them. Remember that a kilometer (km) is 1000 m and a meter is 100 cm.

Lesson 137

Introduction to Sine Waves

GOALS
1. To learn about another part of trigonometry
2. To construct a *sine wave* from a unit circle
3. To see a sine wave with a pendulum

MATERIALS
Worksheet 137
Drawing board, T-square
Clinometer from Lesson 134 or a 15-20 cm string
with a small weight attached, such as a key

ACTIVITIES
Sine wave. If you plot the sine of an angle as it
increases from 0° to 90°, you will get the graph
shown on the right.

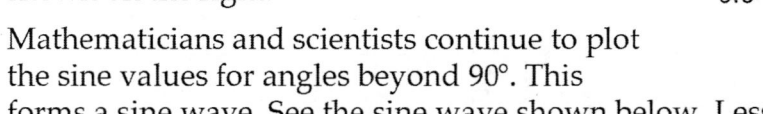

Mathematicians and scientists continue to plot
the sine values for angles beyond 90°. This
forms a sine wave. See the sine wave shown below. Lesson 132 also
has another sine wave.

Constructing a sine wave. One way to construct a sine wave is to
plot the sine of the angle formed by two radii. This is usually done
in a unit circle, so called because it has a radius of 1 unit. The circle
is drawn in an *xy* coordinate system with the center of the circle at
(0, 0). See the first figure below.

One radius lies on the *x*-axis. The other radius starts at the *x*-axis
and moves counterclockwise. On a clock the radius would start at 3
and move backward. When the hand is at the 1, the angle is 60° (see
the second figure). See figures 4-6 for angles greater than 180°.

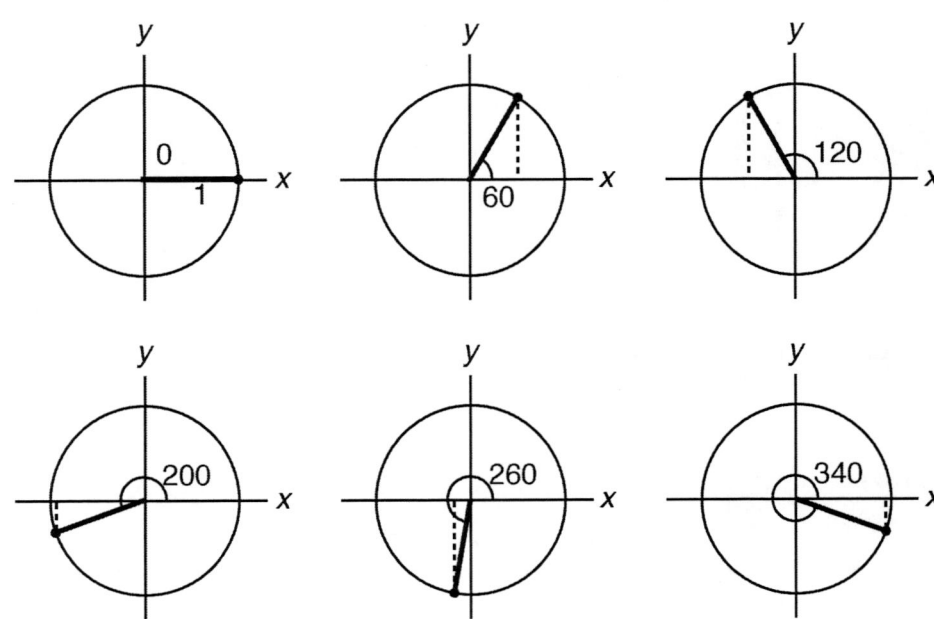

164

As you know, the sine of an angle is the opposite side divided by the hypotenuse. Since the hypotenuse is 1, the sine of an angle is equal to the y value. The dotted lines in the figures are the y values.

Worksheet. For this worksheet, you will construct a sine wave by graphing the y values for every 10 degrees from 0-360. When you have the 37 points, connect the dots to make a smooth sine wave.

As an example, the circles in the figures below show angles for every 30 degrees from 0-360°. By aligning your T-square, you can plot the sine value for the corresponding angle. The left figure below shows plotting the point for 30; the right figure shows plotting the point for 300. Do the worksheet now.

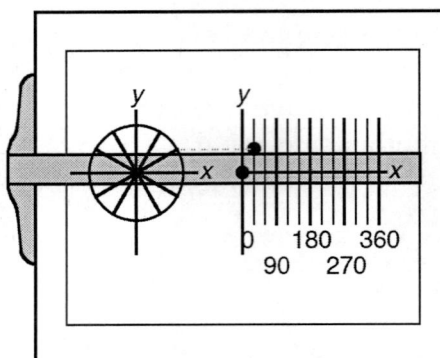 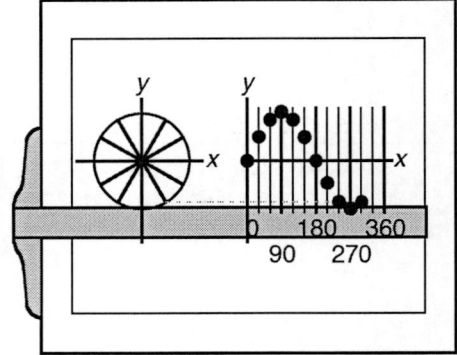

Seeing a sine wave with a pendulum. To see a sine wave formed by a pendulum, you need a light that makes good shadows. The sun is ideal. A pendulum is a weight that swings back and forth. A child's swing is a pendulum. Grandfather clocks also have a pendulum.

Your clinometer makes a good pendulum. Tilt the paper so the weight doesn't touch it when you swing the weight. Now move the clinometer perpendicular to the direction your pendulum is swinging. See the figure on the right. Watch the weight's shadow to see the sine wave.

Waves in nature. Sound waves are sine waves. Waves move energy from place to place. A simple sound source, such as a tuning fork, shown at the left, vibrates like a pendulum. The air molecules move only slightly to transfer the energy to the next molecules of air. Then the molecules return to their original place. Complicated sounds can be broken down into basic sine waves with advanced mathematics.

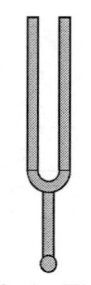

A tuning fork. The "A" note vibrates 440 times a second.

Another example of a sine wave is the wind transferring energy to a body of water. The water moves only slightly to transfer the energy to the next molecules of water. Then the water returns to its original place. The shape of the wave is approximately a sine wave.

Microwaves, radio waves, X-rays, and light waves are all sine waves. The electricity wired in homes in North American is called alternating current, or AC. It has 60 sine waves per second.

Lesson 138

Solids and Polyhedra

GOALS
1. To learn the terms *solid*, *polyhedron* and *polyhedra*, *face*, *edge*
2. To learn expanded meanings for *vertex*, *net*, and *dimension*
3. To construct a tessellation with the geometry panels
4. To construct solids with the geometry panels
5. To find Euler's polyhedron relationship

MATERIALS
Worksheet 138
Geometry panels and rubber bands

ACTIVITIES
Preparing the geometry panels. In your set of geometry panels are four different regular polygons. See the figures below. The numbers tell how many polygons of that shape are in the set.

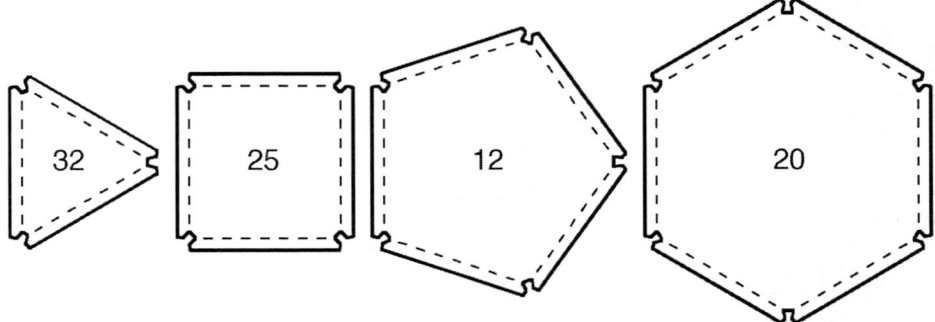

If your set of panels has not been used, you will need to prepare the edges. Bend the perforated (dotted) lines near each edge of the panels. Always bend up toward the colored side. Place the edge on a hard surface and bend gently. It works to do two at a time.

> *Fluorescent lights and being stretched for a long time cause rubber bands to weaken.*

Combining panels. Take two triangle panels and align two edges with the white sides together. Slip a rubber band around the edges as shown on the right. The notches at the vertices keep the rubber bands in place.

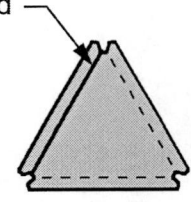
— rubber band —

Solids. A mathematical *solid* has three *dimensions*, depth, or thickness, in addition to width and height. For example, a square has two dimensions while a cube has three dimensions. Likewise, a circle has two dimensions while a sphere has three dimensions. A line has only one dimension because it has neither height nor depth. A point doesn't even have width, so it has no dimensions.

In math, the word "solid" does not refer to being hard, such as a block of wood or chunk of ice.

Polyhedra. When a solid is composed of flat polygons, it is called a *polyhedron* (PAHL-ee-HE-druhn). The *poly* part of the word means "many" and *hedron* means "faces." Usually the plural is spelled "polyhedra," but occasionally as "polyhedrons." Cubes and pyramids are polyhedra while cones and cylinders are not.

The polygons in a polyhedron are called *faces*. The line where the sides of two polygons meet is an *edge*. A *vertex* is the point where three or more edges intersect. Vertices are the "corners" of a polyhedron. See the figure on the right.

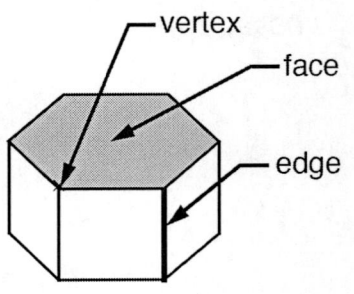

Net. A net of a polyhedron are the polygons drawn flat and connected where possible. You make the polyhedron by folding the figure on the edges. See the figure below. The net of a cube is folded to form a cube.

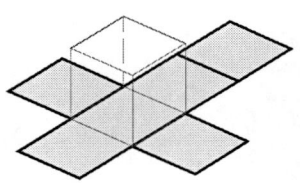 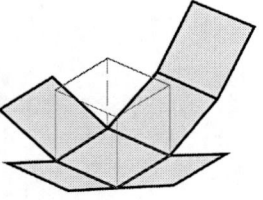 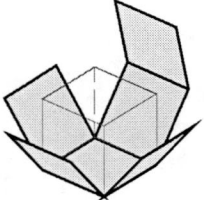 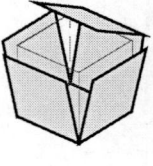

A net being folded into a cube.

A net of a curved solid, such as a cylinder, will have some curved figures. Some solids, such as a sphere, do not have a net. See two examples of nets below.

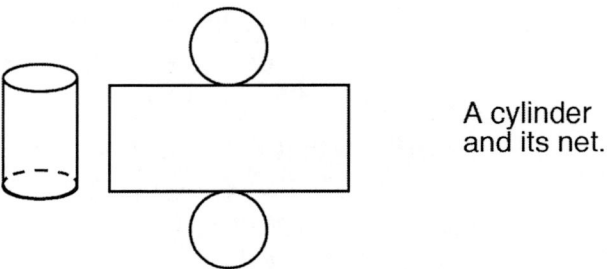

A cylinder and its net.

Problem 1. Make a portion of the tessellation shown on the worksheet. Then, if possible, combine tessellations with your partner and others to make a large design. Before joining several sets, you may want to distinguish sets. Mark the panels in a set on the white side with a distinctive letter or number.

Problems 2-4. The tessellation was a two-dimensional figure. Problem 2 asks you to use part of your tessellation to make a three-dimensional figure, a polyhedron. Start with the net shown and fold it up to make the solid. Record the number of faces, vertices, and edges in the table.

For Problem 3, make several more polyhedra and record the information about them.

Problem 5. There is a special relationship regarding the number of faces, F, vertices, V, and E in a solid. Leonhard Euler discovered it in 1750, although he did not prove it.

Lesson 139

Nets of Cubes

GOALS
1. To visualize a cube from a net
2. To recognize a net of a cube

MATERIALS
Worksheet 139
Geometry panels and rubber bands

ACTIVITIES
Nets of a cube. As you know, a cube has six faces. However, there is more than one net for constructing a cube. There are 11 different nets. Nets that are reflections or rotations of another net are considered the same net.

Not all arrangements of six squares will fold up into a cube, so they are not nets. Use your square panels and find which two arrangements in the figure below are not nets. Also find which two nets are the same. The answers are at the bottom of the page.

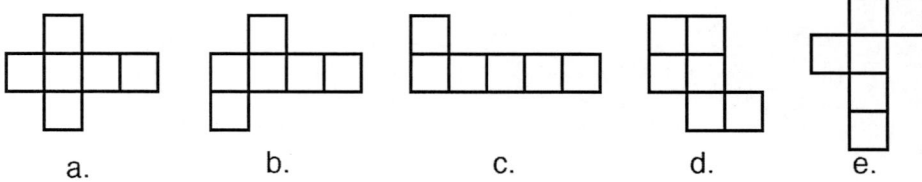

a.　　　　b.　　　　c.　　　　d.　　　　e.

Visualizing aid. To visualize a cube, you might find it helpful to label each square, using the following letters.

F front
L left
R right
T top
B bottom
A aft (back)

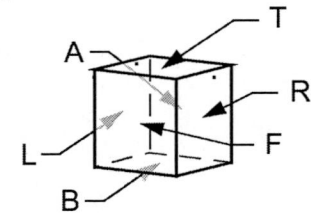

First choose any square for a particular position. In the example below, the square selected for the bottom is marked with B. By mentally folding the panels, fill in the adjacent squares with L and R. Then complete the letters for the remaining squares.

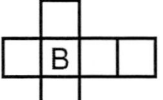 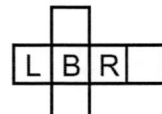 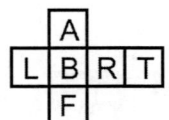 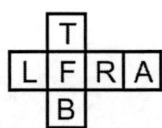

You could also have chosen F (front) as the beginning square as shown in the right figure above.

Notice what happens when you do this procedure with arrangement (d) at the top of the page. You end up with two tops. In other words, you can't make a cube.

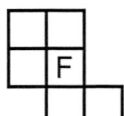 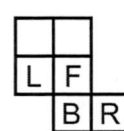 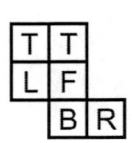

Worksheet. You are to find the 11 nets for a cube among the 40 cube arrangements. [Answers: *c* and *d* are not nets; *b* and *e* are the same.]

Lesson 140

Volume of Cubes

GOALS
1. To learn the terms *volume, cubic centimeter,* and *surface area*
2. To calculate volumes and surface areas of cubes

MATERIALS
Worksheet 140
Geometry panels and rubber bands
Centimeter cubes
Ruler

ACTIVITIES

Volume. *Volume* is the amount of space taken up by a solid. To measure this space you need a unit that takes up three-dimensional space. Units of volume are called cubic units because they are usually cubes.

Recall that you can measure a line segment with centimeters (cm) and area with square centimeters (cm^2). You will measure the volume of a solid with *cubic centimeters (cm^3)*. A cubic centimeter is a cube with edges 1 cm long. See these units of measurements below.

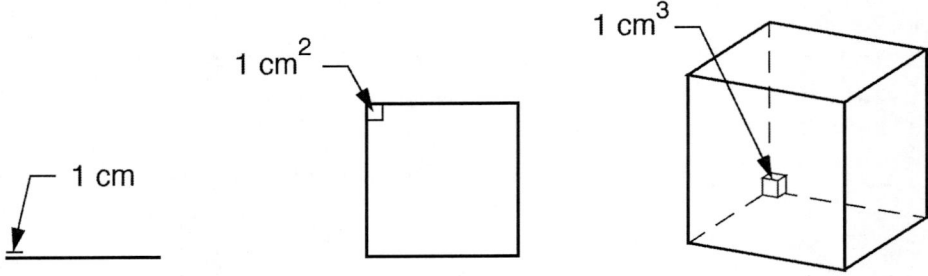

Construct a cube with the geometry panels.

Volume of cubes. Use the centimeter cubes to construct a cube that measures 2 cm on a side. (If you don't have these cubes, look at the figures on the worksheet.) How many centimeter cubes do you need? The number of cubic centimeters you need to fill the solid is the volume. What is the cube's volume? The answers are at the bottom of the page.

Cubing a number. With the cube you made, did you notice that the total number of cubes is 2×2×2? You can also write this expression with exponents as 2^3. Read it as "two cubed."

Surface area of a cube. *Surface area* is the area of all the surfaces of a solid. For polyhedra, the surface area is the sum of the area of all the polygons. The symbol for surface area is "S."

Worksheet. For the first table, you will be considering cubes of various dimensions. The second table pertains to a cube made with the panels, but measured with different units. Measure the edges of a panel polygon as shown in the figure on the right. For the inch, use the nearest whole inch. [Answers: 8, 8 cm^3]

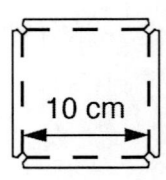

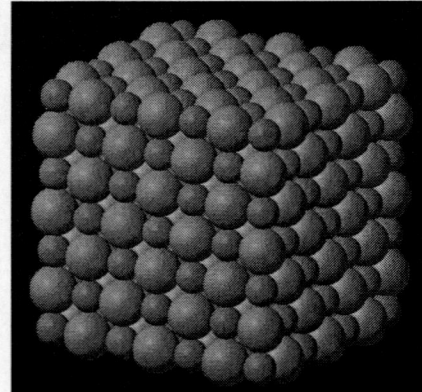

The crystalline structure of common salt. The larger chloride ions form a cubic shape. The smaller sodium ions fill in the gaps between them.

G: © Activities for Learning, Inc. 2010

Lesson 141

Volume of Boxes

GOALS
1. To learn the term *decimeter* and its abbreviation *dm*
2. To learn the term *cube root*
3. To calculate the volume of boxes

MATERIALS
Worksheets 141-1, 141-2
Geometry panels and rubber bands
Drawing board, T-square, 45 triangle
Scientific calculator
Ruler

ACTIVITIES
Decimeters. The length equal to 10 centimeters has its own name, *decimeter* (dess-i-ME-ter). The "deci-" part of decimeter means one-tenth. Since there are 100 cm in a meter, 1 dm is one-tenth of a meter, just as 1 cm is one-hundredth of a meter. The abbreviation for decimeter is *dm* — without any period.

The volume of the larger can is twice the volume of the smaller can.

The edge of a geometry panel is 1 dm. The area of a square panel is 1 dm^2 and the volume of a cube made with the panels is 1 dm^3.

Volume of a box. Construct a cube and a "box" with the panels as shown below. The cube will be 1 dm on a side. The box will consist of two decimeter-cubes for the base and two more cubes on top. Finding the volume of the box in cubic decimeters is the same as finding how many decimeter cubes will fit in the box.

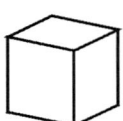 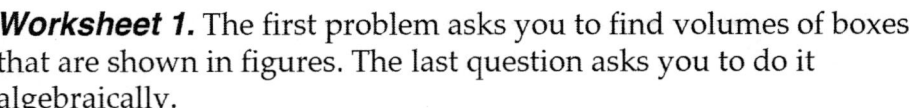

A good way to think of finding volume is to visualize covering the area of the base with cubic units. Then multiply by the number of layers.

A building with cubes in Michigan.

Worksheet 1. The first problem asks you to find volumes of boxes that are shown in figures. The last question asks you to do it algebraically.

Questions 2–8 ask you to find volumes and surface area for boxes without any figures. Watch the units. You will need to convert between centimeters, decimeters, and meters. Do the first worksheet before reading the next section.

Finding cubes and cube roots. On your scientific calculator, you can find cubes and *cube roots*. As you know, the square root of a number is the inverse of a number squared. In the same way, the cube root of a number is the inverse of a number cubed. For example, since $2^3 = 8$, the cube root of 8 is 2. It is written as $\sqrt[3]{8}$.

To find 4^3 on your scientific calculator, press 4 $\boxed{x^3}$ $\boxed{=}$, which gives 64. The *cube root* of 64 is 4. Find it on the calculator by pressing 64 $\boxed{\text{SHIFT}}$ $\boxed{\sqrt[3]{y}}$ $\boxed{=}$.

Worksheet 2. The second worksheet asks for comparisons and drawing a net.

Lesson 142

Volume of Prisms

GOALS 1. To learn the term *prism* and the names of several prisms
2. To calculate volumes and surface areas of prisms

MATERIALS Geometry panels and rubber bands
Worksheet 142
Scientific calculator

ACTIVITIES ***Prism.*** A *prism* is a polyhedron with two congruent parallel polygons with connecting parallelograms. The identical polygons are called *bases*. The remaining polygons are *lateral* faces. The shape of the base determines the name of the prism. See the figures below with the names of several prisms. For the rectangular prism, there are three sets of rectangles that could be the bases.

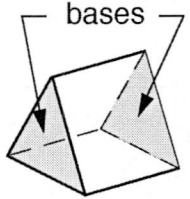

triangular prism

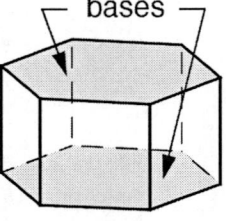
hexagonal prism

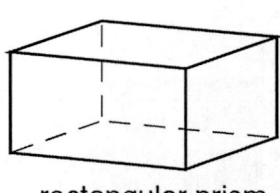
rectangular prism

Notice that the boxes from the previous lesson are now called by their geometrical name, *rectangular prism*.

If the bases of the prism are at right angles to the other faces, the polyhedron is a *right prism*. Otherwise, it is an *oblique prism*. See the prisms on the right.

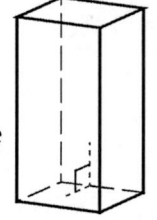
right prism   oblique prism

Is a cube a prism? Is a cube a right prism? Is a cube a rectangular prism? The answers are at the bottom of the page.

Problem 1. After you construct the four prisms, fill in the table. Since you know the volume of a panel cube, the volumes of the first two prisms should be easy. Compare prism C with prisms A and B to make a estimate. Likewise, compare prism D to prism C.

The depth, or height, of a prism is a measure of how tall it is. The letter, *H*, is usually used to show this dimension. Textbooks often use the lower case *h*. The capital *H* is used here to distinguish the height in a polygon from the height of a solid.

Most of the calculations for this problem you can do in your head.

Problem 2. This is the important result of this lesson. If you have a mental image of a prism and understand what volume is, the general volume formula should make sense to you. [Answers: yes, yes, yes]

Lesson 143

Diagonals in a Rectangular Prism

GOALS
1. To learn the terms *short diagonal* and *long diagonal*
2. To calculate the length of diagonals in a prism

MATERIALS
Worksheets 143-1, 143-2
Geometry panels and rubber bands
Drawing board, T-square, 45 triangle
Ruler

ACTIVITIES
Polyhedron diagonals. A *short diagonal* in a polyhedron is a line segment between two nonadjacent vertices on the same face. (Nonadjacent means not next to each other.) A *long diagonal* in a polyhedron is a line segment between two nonadjacent vertices on different faces. Sometimes a long diagonal is called a *space diagonal*. See the examples below for both types of diagonals in a rectangular prism.

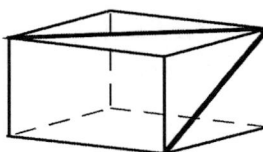

short diagonals

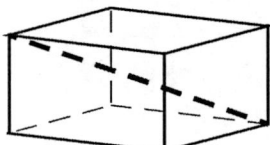

long diagonal

Long diagonal. You can use the Pythagorean theorem for short diagonals. Long diagonals are harder to calculate because they're harder to see. To better visualize the long diagonal, construct a partial cube with a "hinged" top and front. See the left figure below.

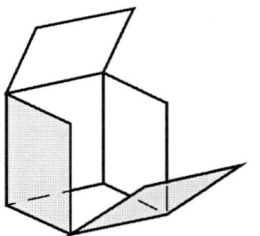

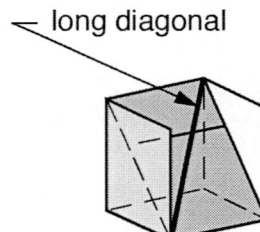

← long diagonal

Now overlap the top and front panels, which divides the cube diagonally into two equal parts. See the right figure above.

Worksheets. In Problem 2, *exactly* means keeping the answer in square root form. If you need to review this procedure, refer to Lesson 60. For Problem 4, use the exact value of the short diagonal to find the long diagonal.

*3-D Pythagorean theorem. You might have noticed after completing the worksheets that there is a formula for long diagonals:

$$D^2 = a^2 + b^2 + c^2$$

Use this formula only if it makes sense to you.

Lesson 144

Cylinders

GOALS   1. To review the term *cylinder*
2. To calculate the volume and surface area of cylinders

MATERIALS   Worksheets 144-1, 144-2
5 squares from the geometry panels, optional
Drawing board, T-square, 45 triangle
A roll of removable tape
Ruler and mmArc compass
Empty food can, (around 440 g (1 lb), also needed for next lesson
Two pieces of paper, standard size

ACTIVITIES   ***Cylinder.*** The left figures below are *cylinders*. Even though a cylinder is a solid with two parallel bases, it is not a prism. The bases, usually circles, are curved, and thus, not polygons. The lateral surfaces are rectangles.

The net for the second cylinder shown below is on the right. The circles can be placed anywhere along side of the rectangle.

A building with many columns.

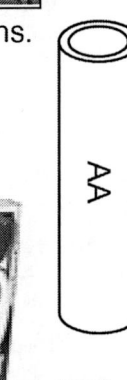

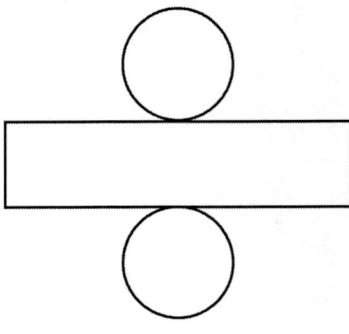

The volume of the larger container is double the volume of the smaller container.

Volume of a cylinder. Finding the volume of cylinder is similar to finding the volume of a prism. First calculate the area of the base (the circle) and then multiply by the height.

Surface area of a cylinder. The surface area involves the two bases, as usual, plus the area of the rectangle forming the side. The width of the rectangle is the circumference of a base.

Worksheets. Problem 1 may be easier to visualize if you make a cube without the top.

For Problem 2, draw a sketch and label what you know.

If you don't have a can, or tin, use a height of 11.3 cm and a diameter of 7.3 cm for Problem 4.

For Problem 5, don't forget the top and bottom of the cylinder.

In Problem 6, to convert from cubic centimeters to cubic decimeters, think how many of one is in the other.

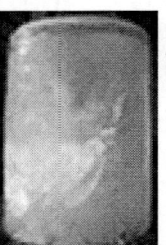
A frozen cylinder of ice in Minnesota.

This organ is made up of many cylinders

Round Tower in Copenhagen, Denmark.

Lesson 145

Cones

GOALS
1. To review the term *cone*
2. To learn the relationship between the volume of a cone and the volume of the circumscribed cylinder

MATERIALS
Worksheets 145-1, 145-2
Scissors, goniometer, ruler, plain sheet of paper
Empty can from the previous lesson
Tape (not removable tape)
Access to tap water or a pitcher with .5 l (1 pint) of water

ACTIVITIES

Cone. When you mention cone, often people think of an ice cream cone. Not all ice cream "cones" are true cones because a cone must have a point.

Worksheet 1. Cut out the rectangle and the sector on the first worksheet. Place two or three pieces of tape on the short edge of the rectangle. Extend half of each piece of tape beyond the edge as shown below. Then fold the rectangle into a cylinder.

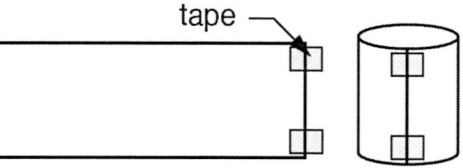

Place tape on one of the radii of the sector. Then fold the sector carefully into a cone. See the right figures above.

Problems 2–5. You will compare the volume of a cone and the volume of a cylinder with the same height and base. In Question 2, write your guess for the relationship. The mathematical proof of this ratio requires calculus. Therefore, you will find the relationship experimentally using water.

First you will construct a paper cone to fit into the empty can. The cone must have the same base and same height as the can. Question 3 guides you through the process. Overlapping on the cone is necessary to make it waterproof. Removable tape will not work when it is wet.

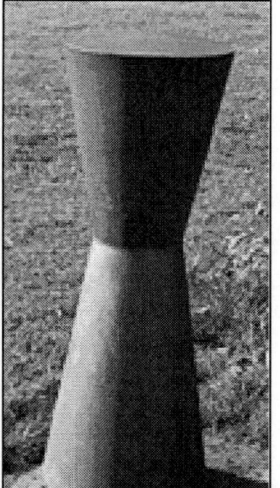

A cone shaped sculpture in Gateshead, England.

Angle of repose. When a solid material such as salt, corn, or wheat is poured into a pile, it forms a cone. The *angle of repose* is the angle of the side of the cone with the surface. See the figure at the right.

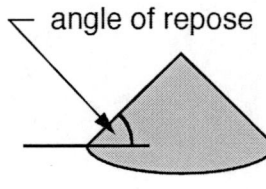
angle of repose

Problem 6. This problem has several steps. To solve such problems, first read the problem several times. Be sure you understand what is being asked. Then draw a sketch and put in what you know. Label what you need to find. This particular problem has the sketches, but not the labels.

You need to use trigonometry to find the pile height. It is a helpful to save your first answer in memory because you will need it again.

Cone shaped trees in Le Havre, France.

Ask others. Show your cylinder and cone to other people. Ask them how many times the cone would fit into the cylinder.

Lesson 146

Pyramids

GOALS
1. To learn the terms *apex*, *regular pyramid*, and *right pyramid*
2. To find the surface area and volume of regular pyramids

MATERIALS
Worksheets 146-1, 146-2
Geometry panels
Calculator and ruler

ACTIVITIES
Pyramid. As you know a pyramid is a polyhedron with a polygon as a base and triangular sides. The sides must meet at a point, called the *apex* (AY-PEX). In a *right pyramid* the line joining the apex and the centroid (center) of the base is perpendicular. A *regular pyramid* is a right pyramid whose base is a regular polygon. The three pyramids below are all regular pyramids.

A pyramid-shaped flower.

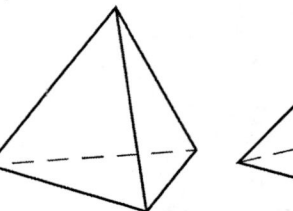

tetrahedron

square pyramid

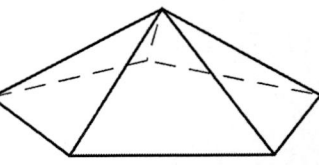

pentagonal pyramid

Construct these pyramids with the geometry panels. You need them for the worksheets. Pronounce pentagonal as (pen-TAH-guhn-ul).

Below are two pyramids that are not regular pyramids. Explain why they are not regular. The answers are at the bottom of the page.

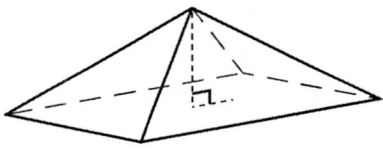

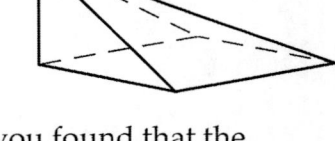

Caldea Spa in Andorra la Vella, Andorra.

Volume of a pyramid. In the last lesson, you found that the volume of a cone is one-third of a cylinder with the same base and height. Pyramids are the same story. The volume is one-third of a prism with the same base and height.

Worksheets. On each worksheet, you will be drawing a net and calculating surface area and volume for a pyramid. To find the volume, you will need to calculate the pyramid's height. To verify your work, you could check your calculations by measuring. Be sure your centimeter answers are accurate to the nearest whole centimeter. [Answers: In the left pyramid, the base is not a regular polygon. In the right pyramid, the line connecting the apex to the centroid is not perpendicular.]

The Pyramid Arena in Memphis, Tennessee.

Square pyramid at Louvre Museum, Paris, France.

Lesson 147

Polygons 'n Polyhedra

GOALS
1. To draw the hexagon that touches six edges of a cube
2. To draw the nets of some special pyramids that fit into a cube
3. To draw the square that touches four edges of a tetrahedron and the half-tetrahedron

MATERIALS
Worksheets 147-1, 147-2, 147-3, 147-4
Geometry panels
Drawing board, T-square, either triangle
Calculator, ruler, and scissors

ACTIVITIES
Worksheet 1. You probably think of squares and right angles when you think of cubes. Amazingly, a hexagon will fit inside a cube. Each vertex of the hexagon touches the midpoint of one of the cube's edges. See the left figure below. On this worksheet, you will construct the hexagon, cut it out and see that it fits.

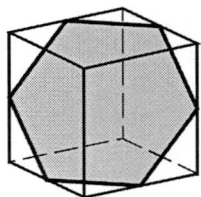 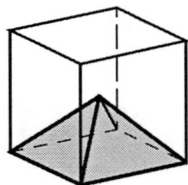 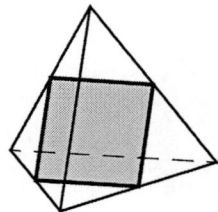

An equilateral triangle will also fit. You will find an interesting relationship between the perimeters of the hexagon and the triangle.

Worksheet 2. The two quadrilaterals on the worksheet will divide a 10-cm cube into identical pyramids. Notice that $\overline{AC}$ is the length of a side of the cube. Also, $\overline{AB}$ is the short diagonal and $\overline{BC}$ is the long diagonal. The pyramids are not right pyramids. For finding volume, remember that the height is the perpendicular distance from the base to the apex.

Worksheet 3. The net for this pyramid (see middle figure above) will not fit on the paper. So you are to make the net with only the four triangles.

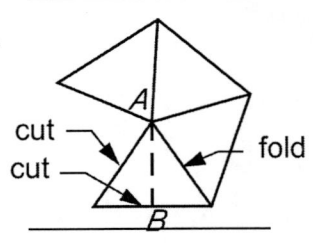

First draw the bottom triangle. You know the base and you can calculate $\overline{AB}$. The easiest way to make the other triangles is by folding and cutting several times. See the figure on the right.

Combine pyramids with others'. It is interesting to take the needed number of pyramids to actually fill a cube. This activity should further convince you that the formula for the volume of a pyramid is one-third the area of the base times the height.

Worksheet 4. You probably think of equilateral triangles and 60° angles when you think of a tetrahedron. But a square fits inside. Each vertex of the square touches the midpoint of one of the tetrahedron's edges. See the right figure above.

Two half-tetrahedra makes a nice puzzle: Can you make a pyramid with them? Can you fit them into your tetrahedron?

A pyramid shaped flower trellis.

Lesson 148

Tetrahedron in a Cube

GOALS
1. To draw the tetrahedron that fits inside a cube
2. To draw the nets of the pyramids that fill the remaining space
3. To find the volumes of all the pyramids
4. To find the volumes in a cube whose side is half the length

MATERIALS
Worksheets 148-1, 148-2, 148-3
Geometry panels
Drawing board, T-square, 30-60 triangle, 45 triangle
2 sheets of plain paper
Calculator, ruler, and scissors

ACTIVITIES
Tetrahedron in a cube. In the previous lesson, you constructed identical square pyramids that filled a cube. Six of them equaled the volume of the cube.

In this lesson, you will construct two different sized pyramids that fill the cube. The large pyramid is a tetrahedron, which touches four vertices of the cube. See the left figure below.

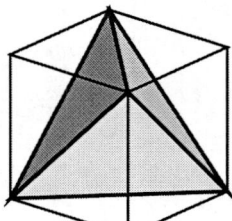 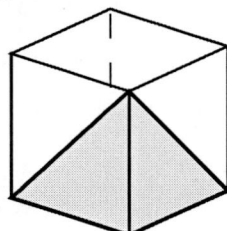

In the space between a face of the tetrahedron and the corner of the cube fits a smaller pyramid. One is shown in the right figure above. You will construct both pyramids and find their volumes.

Worksheet 1. On this worksheet, you are asked to draw the net for the tetrahedron. First you will need to find the length of an edge. The paper will be too small for the entire net. Draw half of the tetrahedron on Worksheet 1 and the other half on a plain sheet of paper. If you have a partner, you can combine triangles.

Worksheet 2. For this worksheet, draw the net for the pyramid that fits into a corner. Plan carefully so it will fit on your paper. Make extras, or combine with a partner, if you want to fill the four corners of a cube.

Problem 3. There is an easy way and a hard way to find the volume of the corner pyramids. Hint: you have more than one choice for the base.

Problem 6. Again, there is an easy way and a hard way to find the volume of the tetrahedron. Refer to Problem 5.

Problem 8. This is an important result. Mentally, fill the large cube with the correct number of smaller cubes to check your answer.

Lesson 149

Platonic Solids

GOALS
1. To learn the terms *Platonic solids* and *dihedral angle*
2. To discover the number of Platonic solids
3. To construct the Platonic solids
4. To practice estimating volumes

MATERIALS
Worksheet 149
Geometry panels
Calculator

ACTIVITIES
Regular polyhedra. If a polyhedron is made entirely of regular polygons and has identical vertices, it is a regular polyhedron. Such solids are called *Platonic solids.* They are named after Plato, who described them in 350 BC. About 1000 years before Plato, people in Scotland carved a set of them out of stone, 3" tall. They are in a museum in Oxford, England. Also see some interesting ones at http://www.rightstartgeometry.com.

Squares into polyhedra. The following activity will help you to understand how many squares are possible at a vertex. Lay out three squares from the panels as shown below with the colored side down. Connect the two edges with rubber bands. Next fold them so they make a vertex. What is the sum of the angles at the vertex? The answer is at the bottom of the page.

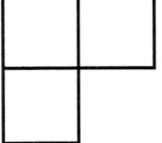 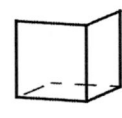 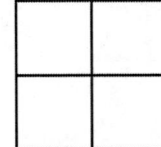

Add another square as shown on the right. Now it is impossible to make a vertex with the four squares. What is the sum of the angles at the vertex? The answer is at the bottom of the page.

Triangles into polyhedra. Lay out three triangle panels. They will make the vertex that is the same as the tetrahedron. Add another triangle. Will that make a polyhedron? Can you add more?

Questions 1-3. Continue adding triangle panels until you have found all possible vertices. Repeat for a pentagon and hexagon. Fill in the table and answer Questions 2 and 3 before reading any farther. [Answers: 270°; 360°]

tetrahedron

cube

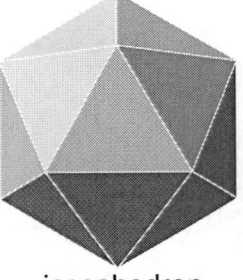
octahedron

icosahedron

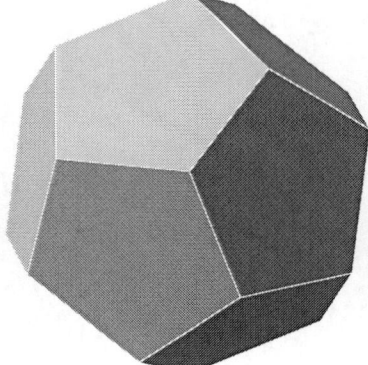
dodecahedron

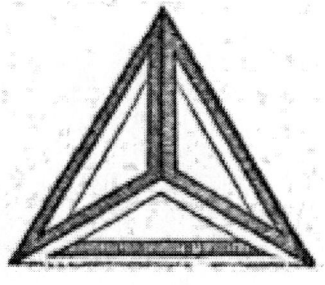

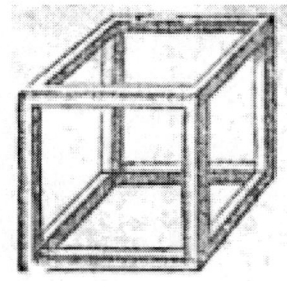

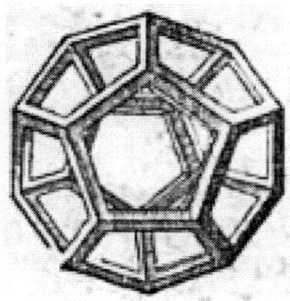

These Platonic solids drawings are from a book written in Latin in 1627.

Platonic solids. Construct the five Platonic solids. You are familiar with two of them, the tetrahedron and the cube. A sketch and name of the other three regular polyhedra are given on the previous page. *Octahedron* (OC-tuh-HE-druhn) has *octa* meaning eight. *Dodecahedron* (DOE-deck-uh-HE-drun) starts with *dodeca-* meaning "two and ten." *Icosahedron* (EYE-cos-uh-HE-druhn) includes the Greek word for twenty, *icosa*. Save these for the next lesson.

Following are some of the characteristics of the Platonic solids.

1. All the faces are congruent polygons.

2. All the *dihedral angles*, the angle between two faces, are equal.

3. The number of polygons at each vertex is the same.

4. Each solid can be inscribed in a sphere. Remember that a regular polygon can be inscribed in a circle.

Questions 3-5. Take a good look at your Platonic models. Especially look for pyramids. This will help you answer these questions. Watch your spelling.

Question 6. This question asks you to verify Euler's theorem for the Platonic solids. See Lesson 138 for reference. Fill in column 7 by substituting into the equation, for example, 3 + 3 = 4 + 2.

Question 7. This a volume question. For column 8 you are to guess the volume of the Platonic solids in cubic decimeters. Compare the volumes with the cube, which you know has a volume of 1 dm³. For the solids less than the cube, think how much of its space would fill the cube. For those solids greater than the cube, think how many times the space in the cube would fit in the solid. You might want to discuss your guesses with a partner.

For the last column, calculate the volumes with the formulas given below the table. Give your answers to two decimal places. How close were you?

Nets of the Platonic solids. You may remember that there are only two nets for the tetrahedron. You also found the 11 nets for a cube. There are also 11 nets for the octagon. The worksheet did not ask you to draw a net for the icosahedron or for the dodecahedron. They each have 43,380 different nets! That could make it a little hard to check solutions. One net for each Platonic solid is shown below.

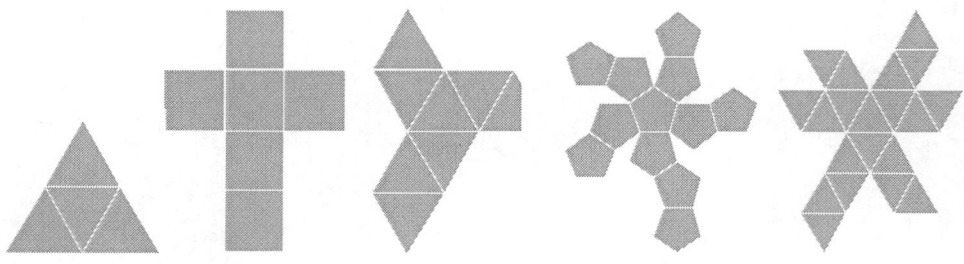

Dodecahedron calendar. Because a dodecahedron has 12 faces, people have made them into a calendar. Each face has one month. There is <u>one</u> at http://www.rightstartgeometry.com.

Lesson 150

Views of the Platonic Solids

GOALS   1. To see the Platonic solids from different views
2. To identify outlines of the Platonic solid and to draw the edges

MATERIALS   The five Platonic solids made from the Geometry Panels
Worksheet 150
Drawing tools

ACTIVITIES   ***Multiple views.*** As you know, solids look different depending upon the way you look at them. Shown below are three views of a tetrahedron. The left figure is probably the easiest to recognize as a tetrahedron. The second figure doesn't necessarily look like a tetrahedron. It could be an equilateral triangle.

The figures directly below have dotted lines to show the hidden edges. Notice how much that helps in visualizing the figures.

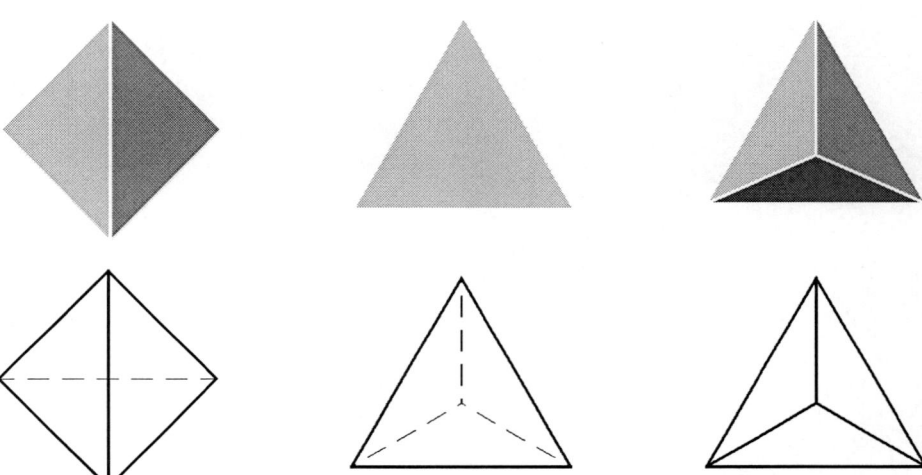

Worksheet. This worksheet will help you appreciate some of the uncommon views of the Platonic solids. First you need to identify which solid it is and then draw the edges. Use your models to help decide where to draw the edges. You needn't draw the hidden edges. Several problems have more than one answer.

Your drawing tools will be helpful for drawing most, but not all, of the lines.

Lesson 151

Duals of the Platonic Solids

GOALS
1. To learn about *dual polyhedra*
2. To draw the duals of some of the Platonic solids
3. To discover some dual relationships

MATERIALS
The five Platonic solids made from the Geometry Panels
Worksheet 151
Colored pencils, optional

ACTIVITIES
Dual tessellations. Recall that a dual tessellation involves forming new polygons around the vertices of a tessellation. The midpoints of the original polygons become the vertices of the new polygons. If you want to review dual tessellations, refer to Lesson 104.

Dual polyhedra. A *dual polyhedron* results when you replace faces with vertices and vertices with faces. Start by making polygons around the vertices of a polyhedron. Do this by connecting the midpoints of the polygons around each vertex of the polyhedron. Each vertex of the dual lies in the midpoint of the face of the polyhedron. Shown below are two Platonic solids with their duals inside the original polyhedra.

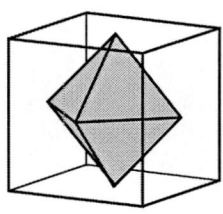 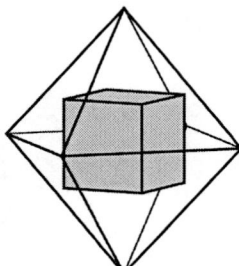

Worksheet. To draw the dual of the cube, connect the midpoints of the squares around the vertex. The midpoints of the visible faces are marked with an "×." The midpoints on the invisible faces are marked with a "+."

To figure out which point is which, it helps to remember how the medians are constructed. The midpoint of a parallelogram is the intersection of the two diagonals. The midpoint of a triangle is the intersection of the medians. See the figures below.

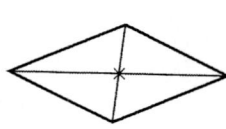 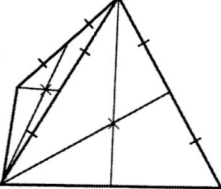

Draw and shade only the faces that are visible. Start with the easiest vertices first. It also helps to look at your models. Do the same for the other three solids. Draw the polygons for the dual by connecting the midpoints of the triangles.

You aren't asked to draw the dual of a dodecahedron. You can figure it out by studying the data in the table and the answers to the questions.

Lesson 152

Surface Area and Volume of Spheres

GOALS
1. To learn the terms *sphere*, *great circle*, and *small circle*
2. To learn to find the volume and surface area of a sphere
3. To solve sphere problems related to volume and surface area

MATERIALS
A ball, optional
Worksheets 152-1, 152-2

ACTIVITIES
Definition of a sphere. Recall the definition of a circle: All the points equidistant from a given point in a plane. Remove the phrase "in a plane" and you have a definition of a *sphere* (SFEER). A sphere is shown at the right.

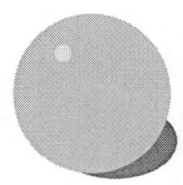

The radius, *r*, of a sphere is the distance from the center to a point on the sphere. The diameter, *d*, is twice the radius.

Great circle. A circle on a sphere that includes the center is called a *great circle*. Look at the figure at the right. As you can see, a great circle divides the sphere into two equal halves. Any other circle on a sphere that doesn't include the center is called a *small circle*.

Volume of a sphere. Think about a sphere in its circumscribed cylinder. See the figure at the right. The volume of the sphere is $\frac{2}{3}$ of the volume of the cylinder. Or, you could say the sphere takes up $\frac{2}{3}$ of the space in the cylinder. Most remarkable!

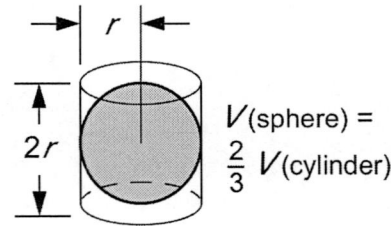

$$V\text{(sphere)} = \frac{2}{3} V\text{(cylinder)}$$

Archimedes considered this to be his greatest discovery. He even asked to have the cylinder and sphere on his tombstone, which was done. The actual formula follows:

$$V\text{(cylinder)} = \pi r^2 h \quad \text{And } h = 2r$$

$$\text{So } V\text{(cylinder)} = \pi r^2 2r = 2\pi r^3$$

$$V\text{(sphere)} = \frac{2}{3} V\text{(cylinder)} = \frac{2}{3} \times 2\pi r^3 = \frac{4}{3}\pi r^3$$

Surface area of a sphere. Archimedes was also delighted about another of his sphere discoveries. He found the surface area of a sphere is $\frac{2}{3}$ the surface area of the cylinder.

Here is a simple way to remember the surface area of a sphere. It is equal to 4 times the area of the great circle. So, assume you have a basketball and 4 pancakes with the same diameter as the basketball. The pancakes will exactly cover the outside of the basketball. But it might be a bit messy.

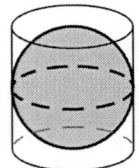 

Surface area of a sphere = Area of 4 great circles

Worksheets. There are questions about these relationships and some calculations.

Lesson 153

Plane Symmetry in Polyhedra

GOALS
1. To learn about *planes of symmetry* in polyhedra
2. To find the planes of symmetry in the Platonic solids
3. To find the planes of symmetry in other polyhedra

MATERIALS
The five Platonic solids made from the Geometry Panels
Worksheets 153-1, 153-2, 153-3
A straightedge, colored pencils (optional)

ACTIVITIES
Plane symmetry. When we talk about mirror symmetry in a square, we show it with a line of symmetry. To show mirror symmetry in a cube, we need a *plane of symmetry.* See the figures below. You can think of a plane as a very large stiff piece of paper.

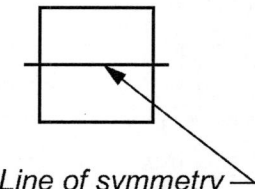
Line of symmetry

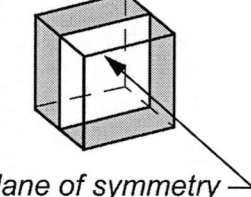
Plane of symmetry

Plane symmetry in a cube. A square has several <u>lines</u> of symmetry and a cube has several <u>planes</u> of symmetry. Study your cube and try to find the nine planes of symmetry. Then look at the figures at the right.

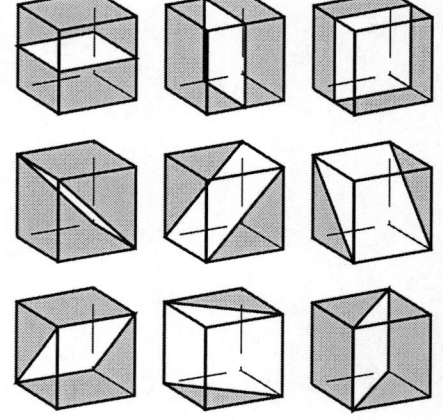

Problems 1–2. Problem 1 has a rectangular prism with square bases. Actually it is two cubes side by side. You are to draw the planes of symmetry.

The rectangular prism in Problem 2 has no regular faces. This means still less symmetry: only three planes of symmetry.

Plane symmetry in a tetrahedron. Look at your tetrahedron model and discover the planes of symmetry. Notice each plane goes through an edge and the opposite face. Since you know the number of edges, you know the number of symmetry planes. Draw these for Problem 3.

Plane symmetry in a regular prism. In a regular prism the diagonals in the base polygons determine symmetry planes. However, there is one more: the plane halfway between the two bases. Problem 4 asks to draw them for a pentagonal prism.

Questions 5-16. These are an assortment of questions.

Summary. Symmetry of solids is very important in fields outside of mathematics. Physics, chemistry, biology, medicine, and minerals all depend upon symmetry. It is also essential for artists and jewelers.

G: © Activities for Learning, Inc. 2010

Lesson 154

Rotating Symmetry in Polyhedra

GOALS
1. To learn about *axes of symmetry*
2. To find the axes of symmetry of the Platonic solids
3. To discover some dual relationships

MATERIALS
The five Platonic solids made from the Geometry Panels
Worksheet 154-1, 154-2

ACTIVITIES

Rotating symmetry in a cube. Take your cube and touch the center of the left side with your left middle finger. Hold the center of the right side with your right middle finger. The dots in the first figure on the right shows the placement. Using your thumbs, rotate your cube about an imaginary line between the two dots.

How many times will the cube be congruent to the starting face? That answer gives the order of rotation. In this case it is 4-fold.

Axes of symmetry. This imaginary line in the cube is called an *axis of symmetry.* The plural of *axis* is *axes.* Axes are related to the point of rotation in polygons and always go through the center of the polyhedron.

A cube has more than one axis. How many axes are there between face centers? The left figure below shows three. There are also four axes between opposite vertices. See the middle figure below. What is their order of rotation? The answer is at the bottom of the page. Lastly, there are six axes between the midpoints of opposite edges. See the right figure.

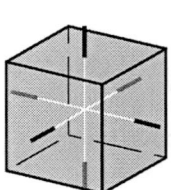 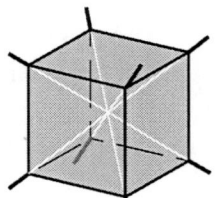 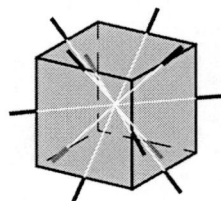

Table A. After finding the axes of symmetry in your cube, complete the first line of Table A on Worksheet 1. Next take the octahedron and check for axes of symmetry through opposite faces, vertices, and edges. Record those results. Also add the total number of axes and record that. Repeat for the dodecahedron and icosahedron.

Table B. The tetrahedron is saved for last because it has only two types of axes. Find them and record it in Table B.

Worksheet 2. These problems take some mental maneuvering. Questions 9 and 10 have amazing answers. [Answers: 3-fold; 2-fold]

Circumscribed Platonic Solids

GOALS
1. To measure and calculate the radii for the spheres that circum-scribe the Platonic solids
2. To find the ratio of the volume of the solid to the volume of the circumscribed sphere
3. To review *reciprocal*

MATERIALS
The five Platonic solids made from the Geometry Panels
T-square, scientific calculator
Worksheet 155

ACTIVITIES
Sphere symmetry. First, think about how many planes of symmetry a sphere possesses. Yes, it is an infinite number.

How many axes of symmetry does a sphere have? The answer is at the bottom of the page. Each axis has what order of rotation. The answer is at the bottom of the page. This makes a sphere perfectly symmetrical.

Spheres and Platonic solids. Consider how much symmetry the Platonic solids have. It should not be too surprising that a sphere circumscribes each Platonic solid. That means that each vertex of the solid lies on the sphere.

Finding the radius of the circumscribed sphere. You can find the approximate radius, R, of the circumscribed sphere by direct measuring. It is a little tricky, but easier with your T-square.

For the cube, first set one vertex on the edge of a surface. Hold it with the opposite vertex directly above it. Use your T-square as shown in the figure at the right and measure the height, which is the diameter.

Worksheet. For this worksheet, you will be comparing the volume of the Platonic solids with the volumes of the circumscribed spheres. First, you will calculate the radius of the solid. It should be close to your measurement. Don't forget you are measuring a diameter, while you need a radius.

While calculating, pay particular attention to square root signs and parentheses. Write down your answer and continue with the volume of the circumscribed sphere. Do not reenter any numbers.

Calculator hints. Notice that you will need to divide another number by your just calculated number. One way to do this is to put your calculated number into memory. Then key in the other number and divide by the number in memory.

There is a special key on scientific calculators for finding the *reciprocal* of a number. (A reciprocal is 1 divided by a number.) For example, the reciprocal of 2 is .5 and the reciprocal of .5 is 2.) The key on your Casio is $\boxed{x^{-1}}$. On some scientific calculators it is $\boxed{1/x}$.
[Answers: infinite; infinite]

Lesson 156

Cubes in a Dodecahedron

GOALS 1. To visualize and construct the cubes in a dodecahedron
2. To find the volume of the cube

MATERIALS Worksheet 156
The dodecahedron made from Geometry Panels
Colored pencils, optional
Calculator

ACTIVITIES ***Cube in a dodecahedron.*** Notice that a dodecahedron has 12 faces and a cube has 12 edges. Suppose we put each edge of a cube on a diagonal of a pentagon. See the figure on the right. The vertices of the cube lie on the vertices of the dodecahedron.

Worksheet. Can the cube fit more than one way inside the dodecahedron? How many diagonals are in a pentagon? That's the number of different ways the cube will fit in the dodecahedron. You are to draw these cubes on the worksheet, one per dodecahedron. Be sure you don't repeat cubes. Choose one pentagon and check the diagonals. Use your model dodecahedron to help you visualize the cubes. Color the cubes if you want.

Question 6. To find the volume of the cube, you need to find the length of the pentagon diagonal. Look at the figures below. Think about how you could calculate that length, *RT*. Then read the following hints.

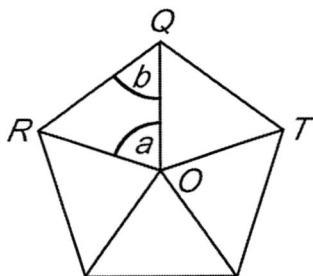 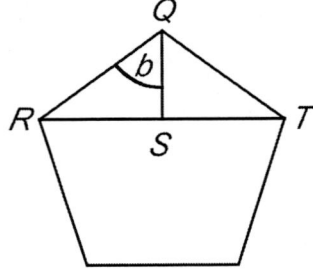

The length you're looking for is *RT*.

If you can find *RS*, you can find *RT*.

If you can find $\angle b$, you can use trig to find *RS* because you know the length of *RQ*. Also $\angle RSQ$ is a right angle. (See Lesson 131.)

If you can find $\angle a$, then you can find $\angle b$, because $\triangle ROQ$ is an isosceles triangle.

You can find $\angle a$ because you know the sum of the angles around a point.

Question 7. Be sure to take a guess before finding the answer. You needn't recalculate the volume of the dodecahedron. You did it in Lesson 149. Do the calculation and then continue reading below.

Bonus. Did you notice the length of the cube is ϕ, the golden ratio?

Lesson 157

Stella Octangula

GOALS
1. To construct the *stella octangula*
2. To view a *concave polyhedron*

MATERIALS
Worksheet 157
The tetrahedron, octahedron, and icosahedron, made previously
Colored pencils, optional

ACTIVITIES

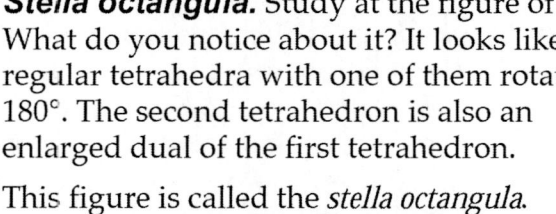

Stella octangula. Study at the figure of the right. What do you notice about it? It looks like two regular tetrahedra with one of them rotated 180°. The second tetrahedron is also an enlarged dual of the first tetrahedron.

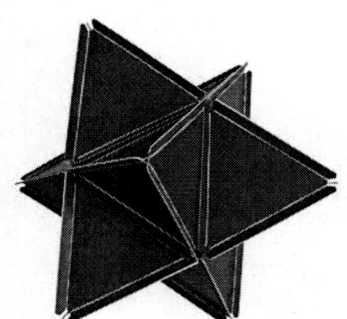

The stella octangula made with the geometry panels.

This figure is called the *stella octangula.* Johannes (yo-HON-is) Kepler (1571-1630) gave it this name although he was not the discoverer. (*Stella* is Latin for star.) Kepler was a German mathematician, astronomer and science fiction writer. Escher made a famous drawing using this shape. See http://www.rightstartgeometry.com.

Building a stella octangula. You can also consider the stella octangula as eight smaller tetrahedra around a core. The core is an octahedron. You could build the stella octangula using triangular panels, but it is difficult. An easier and more interesting way is to build it starting with an octahedron.

Keep your octahedron intact. Make eight 3-sided tetrahedra with the remaining triangles. Attach each tetrahedron to only one edge of the octahedron with a rubber band. See the figure on the right. Then you can peek inside your stella octangula to see the octahedron.

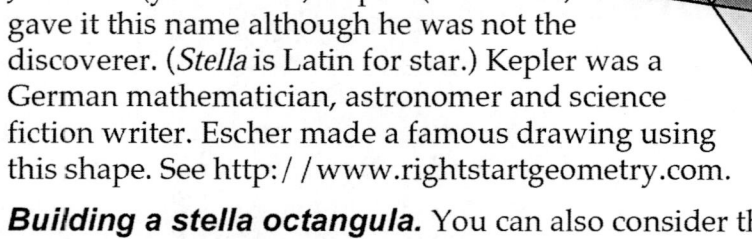

Convex and concave polyhedra. The definitions for convex and concave polyhedra are similar to those for polygons (Lesson 55). A polyhedron is *concave* if any angle between two faces is greater than 180°. The stella octangula is *concave*, while the Platonic solids are convex. To see some beautiful images of additional convex polyhedra, go to http://www.rightstartgeometry.com.

Worksheet. The stella octangula fits into a cube. The edges of the two large tetrahedra are the short diagonals of a cube.

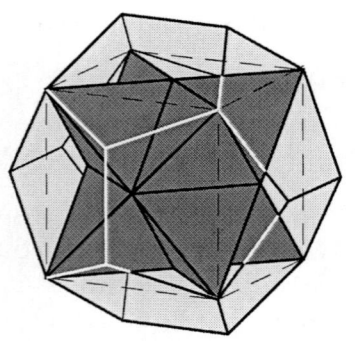

Can you find four Platonic solids in this figure?

If you think you can solve this problem without help, go ahead. If not, continue reading. When you see a problem like this, read it several times to be sure you understand what is asked. Here you are asked to compare two volumes, so you must find both of them.

One way to find the volume of the stella octangula is to find the volume of all the little tetrahedra and the octahedron.

To find the volume of the cube is a little harder. You know the diagonal of the square. So use the Pythagorean theorem to find the length of a side. The ratio is simply the division of the two volumes. Remember to treat $\sqrt{2}$ as a number in your equations.

Lesson 158   **Truncated Tetrahedra**

GOALS   1. To extend the meaning of *truncate*
2. To learn about *semiregular polyhedra* and the *Archimedean solids*
3. To compare lengths, area, and volume of several tetrahedra

MATERIALS   Geometry panels
Worksheet 158
Drawing tools
Calculator

ACTIVITIES   ***A truncated tetrahedron.*** Take four hexagon panels and build the solid shown on the right. Add four triangles to the "open" areas. Study the solid carefully to understand why it is called a *truncated* tetrahedron. Truncated means "chopped off."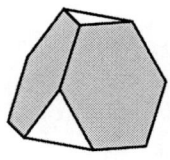

Archimedean solids. There are 13 semiregular polyhedra, called *Archimedean solids*. They are pictured on page 196. Archimedes is considered to be the first person to study and write about them. However, his work is now lost.

Questions 1-4. Next, look at the vertices. Are they all the same? Remember that tessellations having identical vertices with more than one regular polygon are *semiregular*. Likewise, solids with identical vertices, except Platonic solids, are *semiregular polyhedra*.

A set of tetrahedrons. Replace the four triangles of the truncated tetrahedron with small tetrahedra. See the right figure below. Also make a small tetrahedron with four panels as shown on the left. Finally, make a medium-sized tetrahedron as shown in the middle. Skip the bottom for the middle tetrahedron because you will not have enough triangles.

   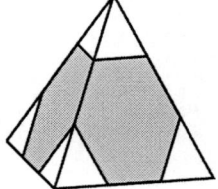

Finding the surface areas. For Question 5, use the area of a triangle panel as the unit of area. Then the area is simply the number of triangles it takes to cover the tetrahedron.

Finding the volumes. For Question 6, you calculated the volume of the small tetrahedron in Lesson 146. Find the volumes of the remaining two tetrahedrons by calculating or by measuring. Find the height, *h*, of the base triangles and the height, *H*, of the pyramids.

Finding the ratios. Use the dimensions and results from Questions 5 and 6 for the table in Question 7. Just use common sense for the *h* and *H* columns. Don't bother with any formulas or algorithms. Continue the patterns to find the data for the row of the "extra large" tetrahedron.

Lesson 159

Truncated Octahedron

GOALS
1. To construct a truncated octahedron
2. To relate the truncated octahedron to the octahedron
3. To find the volume of the truncated octahedron through working with pyramids

MATERIALS
Geometry panels
Worksheet 159
Straightedge
Calculator

ACTIVITIES
A truncated octahedron. Use the net shown below and construct a truncated octahedron. It isn't necessary to add all the squares.

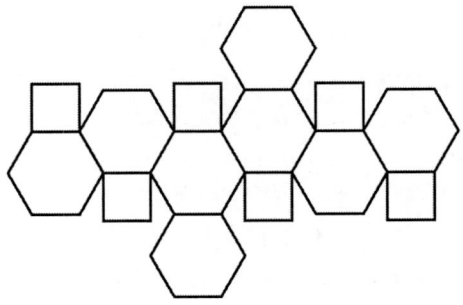 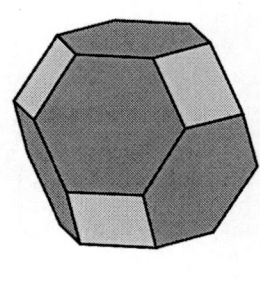

Does your polyhedron look like an octahedron minus its vertices? To see it better make some square pyramids and place them on top of the squares. Also see the figure at the right.

Problem 1. For this problem, you are to draw the octahedron that fits around the truncated octahedron. Look at your model with an attached square pyramid. Notice how the lines of the hexagons continue into the pyramid. In Problem 1 extend the lines of the hexagons in the same way.

Problems 2-4. These problems are all based on finding the area of your truncated octahedron. Finding its volume at first seems like a tough problem. But thinking in terms of pyramids makes it much easier. Remember that an octahedron is two square pyramids attached.

Also recall that the volume of a pyramid is

$$V = \frac{1}{3}BH$$

B is the area of the base and *H* is the height of the pyramid.

For a pyramid whose sides, *a*, are equilateral triangles, *H* is

$$H = \frac{1}{2}\sqrt{2} \times a$$

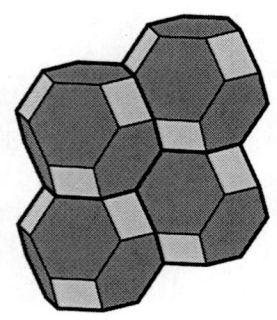

Filling space. If your class has made at least four truncated octahedra, try to put them together without gaps. It is the only Archimedean solid that can fill space.

Lesson 160

Truncated Icosahedron

GOALS 1. To construct the "soccer ball" with the panels
2. To construct a truncated icosahedron from an icosahedron
3. To calculate the number of faces, vertices, and edges in the truncated icosahedron

MATERIALS Worksheets 160
Geometry panels
Straightedge
Colored pencils (optional) and scissors

ACTIVITIES ***Truncated icosahedron.*** You probably are familiar with the truncated icosahedron because it is the shape of the soccer ball. Use the net below and construct the truncated icosahedron. You will need all 20 hexagons and 12 pentagons from your panel set.

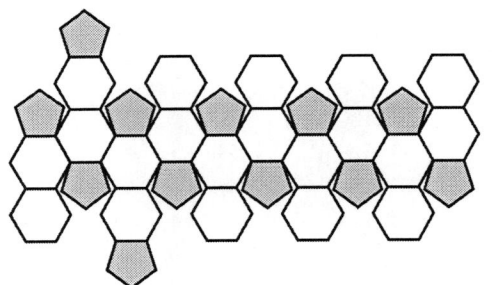 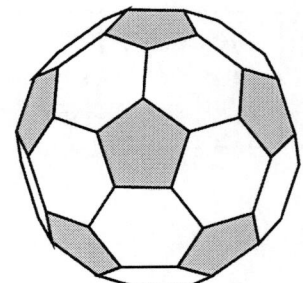

Soccer ball. In the 1950s Danish soccer players started using a soccer ball made in the shape of the truncated icosahedron. It wasn't until 1970 that the first world cup games used this ball. The game was played in Mexico. Two reasons have been given for making the pentagon panels black. Such a ball was easier to see on black and white television. Also, the players could more easily see if the ball were swerving.

Carbon. Carbon is a very common element in the physical world and a part of all life. It is an important part of petroleum and coal. Carbon also exists separately as diamond, the hardest natural substance known, and graphite.

The atoms in carbon arrange themselves in polyhedrons. The atoms in the diamond molecule are arranged as the vertices of a tetrahedron. Graphite is mixed with clay to form the "lead" in pencils. The graphite molecule forms a hexagonal prism.

In 1985 a rare form of carbon was discovered. Because its molecule has 60 atoms, it is called carbon-60. Amazingly, the molecules are at the vertices of a truncated icosahedron.

Problems 1–4. Here you are given an icosahedron. Imagine what shape you will get if you truncate (cut off) each vertex at the points marked by ×'s. (Some ×'s are on invisible lines.)

Draw the truncated icosahedron. Scissors allows you to actually truncate six of the vertices.

Questions 5–11. These questions require some thinking. Often an answer will depend upon the previous questions.

The Eiffel Tower and soccer ball in Paris, France.

Lesson 161

Cuboctahedron

GOALS
1. To learn about the cuboctahedron
2. To see how the cuboctahedron relates to the cube and octahedron

MATERIALS
Worksheets 161–1, 161-1
Drawing board, T-square, and 30-60 triangle
Calculator
Scissors

ACTIVITIES
Cuboctahedon. The name *cuboctahedron* (CUBE-oct-ah-HE-drun) is a compound word made of *cube* and *octahedron.* Usually the "e" in the word *cube* is dropped.

Construct the cuboctahedron from the net given below. Leave one of the squares attached by only one rubber band. This allows you to peer inside.

> **The cuboctahedron will be needed again in the next two lessons.**

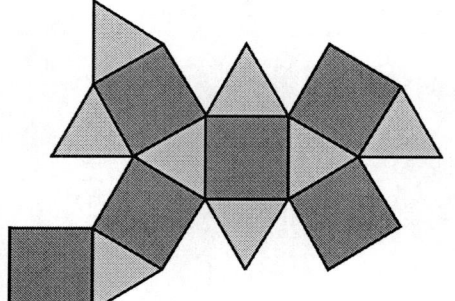 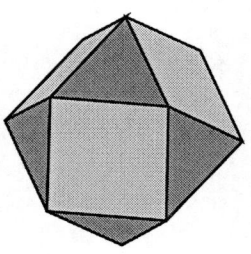

A different way to imagine a cuboctahedron is to think of a cube and an octahedron. Align them as shown at the right and mentally move them through each other. The intersection of the two solids is a cuboctahedron.

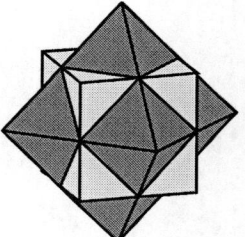

Problems 1–2. Orient your cuboctahedron so you can see how it resembles a cube. This is best done with the squares on top, bottom, directly in front, and on both sides. The first question on the worksheet asks you to draw the cuboctahedron in a cube.

While it is still in that position, concentrate on the triangles. Notice how the four upper triangles could continue to make a square pyramid. The bottom four triangles do the same. The two pyramids form an octahedron.

Problem 3. You can get some interesting views by turning your cuboctahedron to various positions. This may help for this problem.

Problems 4–6. These are volume problems. The rounded answer for Problem 4 isn't accurate enough for Problem 5. One way is to use your calculator answer from Problem 4 to find the volume in Problem 5.

For Problem 6, compare the answer for Problem 5 with the volume of a panel cube.

Problem 7. Peek inside your cuboctahedron. Notice that a hexagon will fit around its center. Draw it, cut it out, and place it inside.

Lesson 162

Rhombicuboctahedron

GOALS
1. To learn about the rhombicuboctahedron and the great rhombicuboctahedron
2. To solve some geometric problems

MATERIALS
Worksheets 162–1, 162-2
Geometry panels, including the cuboctahedron
Calculator

ACTIVITIES
Rhombicuboctahedon and great rhombicuboctahedron. The rhombicuboctahedron (rom-bi-CUBE-oct-ah-HE-drun) sometimes is called the *small rhombicuboctahedron.* This distinguishes it from the great rhombicuboctahedron. The left figure below is the rhombicuboctahedron; the middle figure is the great rhombicuboctahedron.

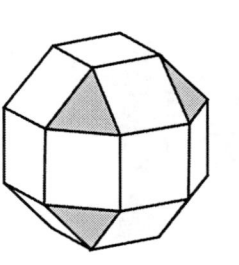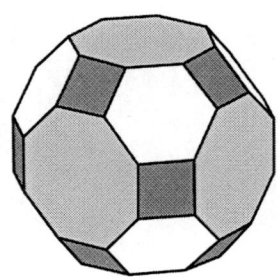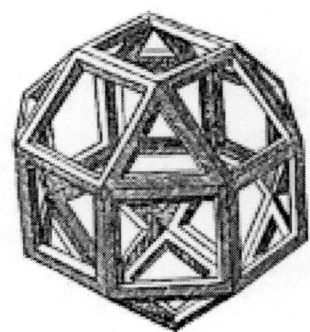

Leonardo da Vinci in 1439 drew an open rhombicuboctahedron. See it above on the right. For more of da Vinci's drawings, go to http://www.rightstartgeometry.com.

Rhombicuboctahedron.
The net for the rhombicuboctahedron is shown at the right. Construct it with your geometry panels.

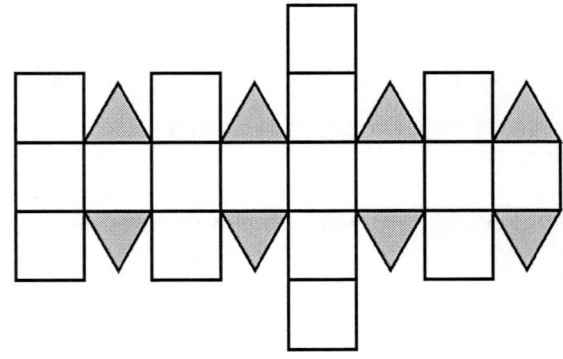

Then compare it to your cuboctahedron. See how they are both related to the cube.

Worksheet 1. Your rhombicuboctahedron model will help you answer Questions 1–8. For Question 5, you needn't count the edges. Figure out the total number of edges in all the faces. But remember two polygons join to make one polyhedron edge.

You can answer Question 6 in the same way. You know the total number of vertices for all the faces. Then consider how many vertices join at each polyhedron vertex.

Worksheet 2. These problems are all based on the smallest cube that the rhombicuboctahedron will fit into. Read each problem at least twice. Be sure your answers make sense.

Lesson 163

Icosidodecahedron

GOALS
1. To learn about the icosidodecahedron
2. To learn about *quasiregular* polyhedra

MATERIALS
Worksheets 163-1, 163-2
Geometry panels, including the cuboctahedron
Straightedge, ruler
Colored pencils, optional
Calculator, optional

ACTIVITIES
Icosidodecahedron. Construct the icosidodecahedron
(eye-cos-i-DOE-deck-uh-HE-drun) from the net given below. It may
be easier to make if you notice that every edge has a pentagon and a
triangle. Leave one or two pentagons attached by one rubber band.

The figures are on the right
show truncating the corners
of the dodecahedron. First it
morphs into a truncated
dodecahedron and then an
icosidodecahedron.

Quasiregular polyhedra. A *quasiregular* polyhedron has only two
kinds of regular faces. They alternate around each vertex. Each
polygon is entirely surrounded by the other type of polygon. Study
the cuboctahedron and the icosidodecahedron carefully. They are
both quasiregular.

Recall that the cuboctahedron can be split into two equal parts along
the edges. The resulting base is a hexagon. See the left two figures
below. This base is similar to a "great circle" and is often called that.
An icosidodecahedron also has a great circle. See the right two
figures. Great circles are a property of quasiregular polyhedrons.

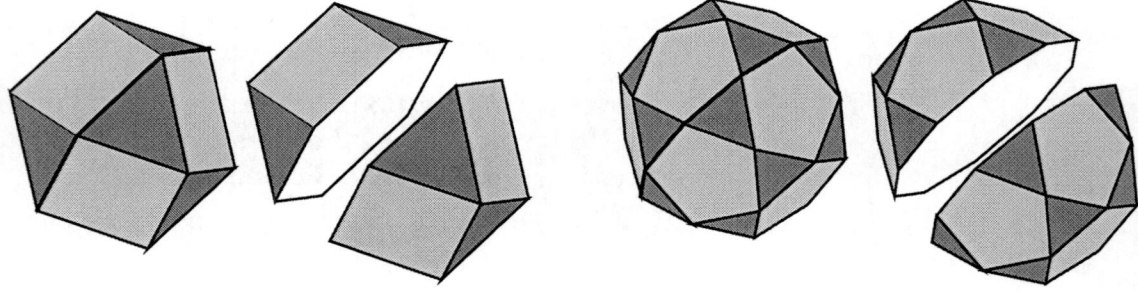

Half of the cuboctahedron is another polyhedron called a triangular
cupola. Half of the icosidodecahedron is a pentagonal rotunda.

Lesson 164

Snub Polyhedra

GOAL 1. To learn about the snub cube and the snub dodecahedron

MATERIALS Worksheet 164
Geometry panels
Drawing board and T-square

ACTIVITIES ***Snub polyhedra.*** Among the 13 Archimedean solids are the snub cube and snub dodecahedron. They are formed by adding extra triangles around the squares of a cube or the pentagons of a dodecahedron.

Snub cube. This activity is best done with two people, each having a set of geometry panels.

A snub cube is a cube with extra triangles added around each side and vertex of the square. To make the snub cube, start with the net for a cube shown at the right. Then spread the squares apart.

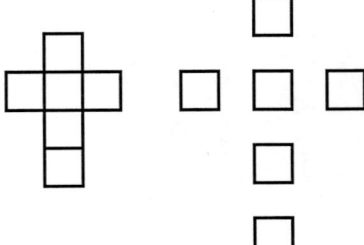

Next rotate each square approximately 17° about its center. One person rotates the six squares in the clockwise direction as shown below.

The other person rotates the squares in the counterclockwise direction as shown below. Both people fill in the spaces with 32 triangles and construct the snub cube.

Notice that the darker equilateral triangles share vertices with three squares and share edges with only other triangles. The lighter triangles share an edge with a square.

When you have completed the snub cubes, look for symmetry. Do you see plane symmetry or axes of symmetry? Compare the left- and right-handed models. How are they related? (If you don't have a partner, compare the figures on the previous page.)

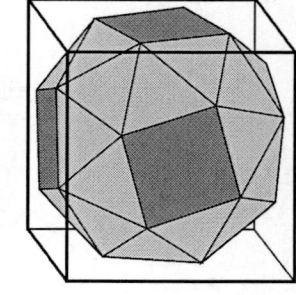

To see how your snub cube would fit into a cube, orient it as shown on the right.

Snub dodecahedron. Like the squares in the cube, the pentagons in the dodecahedron can be separated. The space is also filled with triangles. The panel kit does not include enough triangles to make the snub dodecahedron. The two versions of the snub dodecahedron are shown below.

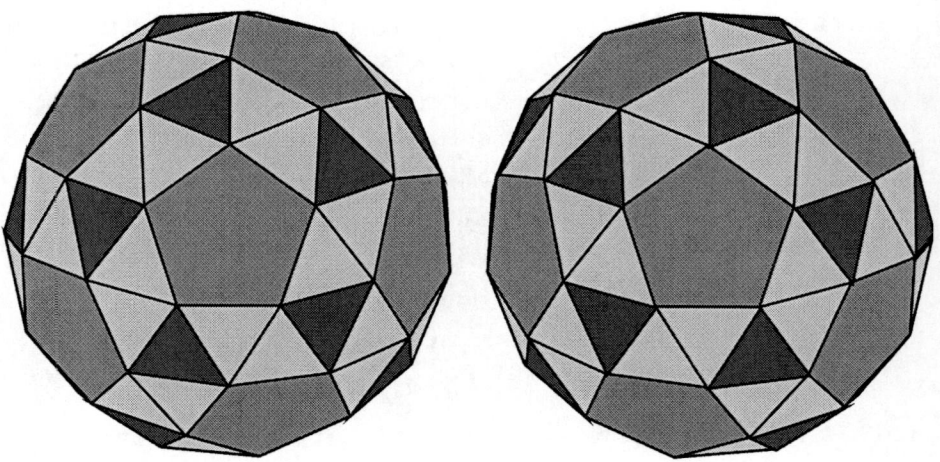

Notice, again, in the figures above that the darker equilateral triangles share vertices with three pentagons. They share edges only with other equilateral triangles. The dark triangles also seem to be in the center of a larger equilateral triangle.

Worksheet. The first six questions deal with snub cubes. Question 7 asks you to draw the triangles to make the other version of the snub dodecahedron. It is harder that it first looks. But it helps to line up and compare vertices with the given model.

Symmetry in the human body. The human body appears quite symmetrical on the outside. Inside, however, is a different story. Most of the heart resides on the body's left side, making the left lung smaller. The left lung has only two lobes while the right lung has three. The spleen and stomach also occupy space on the left side. Located on the right side are the gallbladder, the appendix, and most of the liver.

About 1 out of 10,000 people have *situs inversus,* in which organ placement is reversed. Such a person is a mirror image of someone with the usual organ arrangement. Do you see how this is similar to the two forms of the snub cube or snub dodecahedron?

Lesson 165

Archimedean Solids

GOALS
1. To review the Archimedean solids
2. To find the relationship between the total angles of a polyhedron and the number of vertices

MATERIALS
Worksheets 165-1, 165-2
Geometry panels, optional, and calculator

ACTIVITIES
The Archimedean solids. The following page has figures of the Platonic solids and the Archimedean solids. Notice that each Platonic solid has a truncated version in the Archimedean solids. Try mentally to truncate several of the Platonic solids.

Two of the Archimedean solids have not been previously discussed. They are the rhombicosidodecahedron and the great rhombicosidodecahedron, both shown below.

> **To see Archimedean sculptures:** *http://www. rightstartgeometry.com*

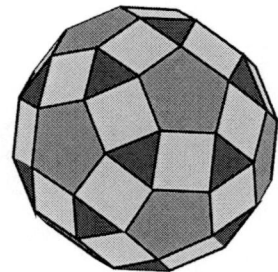 

You could construct the rhombicosidodecahedron if you left openings for five squares. (It needs 30 squares, but the panel sets includes only 25.)

By their names and shapes, you can see they are related to the icosidodecahedron. The icosidodecahedron and the rhombicosidodecahedron both have 20 triangles and 12 pentagons. But the rhombicosidodecahedron has 30 squares. The great rhombicosidodecahedron has 20 hexagons, 12 decagons, and again 30 squares.

Worksheet 1. For this worksheet, you will need to find the vertex code for each Archimedean solid. Also complete the chart for the number of faces, vertices, edges, and individual polygons.

Angles in a polyhedron. Finding the surface angles in a polyhedron means to add the angles in each polygon. In a tetrahedron, it is quite simple. Each triangle has 180°, so the surface angles total $180 \times 4 = 720°$.

There are several ways to find the angles in a pentagon. In the first figure on the right are five triangles. You know the angles in them, but you need to subtract the angles in the center, which total 360°.

Another way is to divide the pentagon into three triangles, as shown on the far right. Of course, if you knew the angle formed at the pentagon vertex, you could simply multiply it by 5.

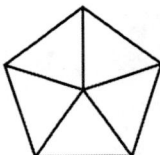 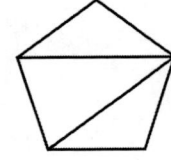

Worksheet 2. Here you find the relationship between the sum of surface angles of a polyhedron and the number of vertices.

196

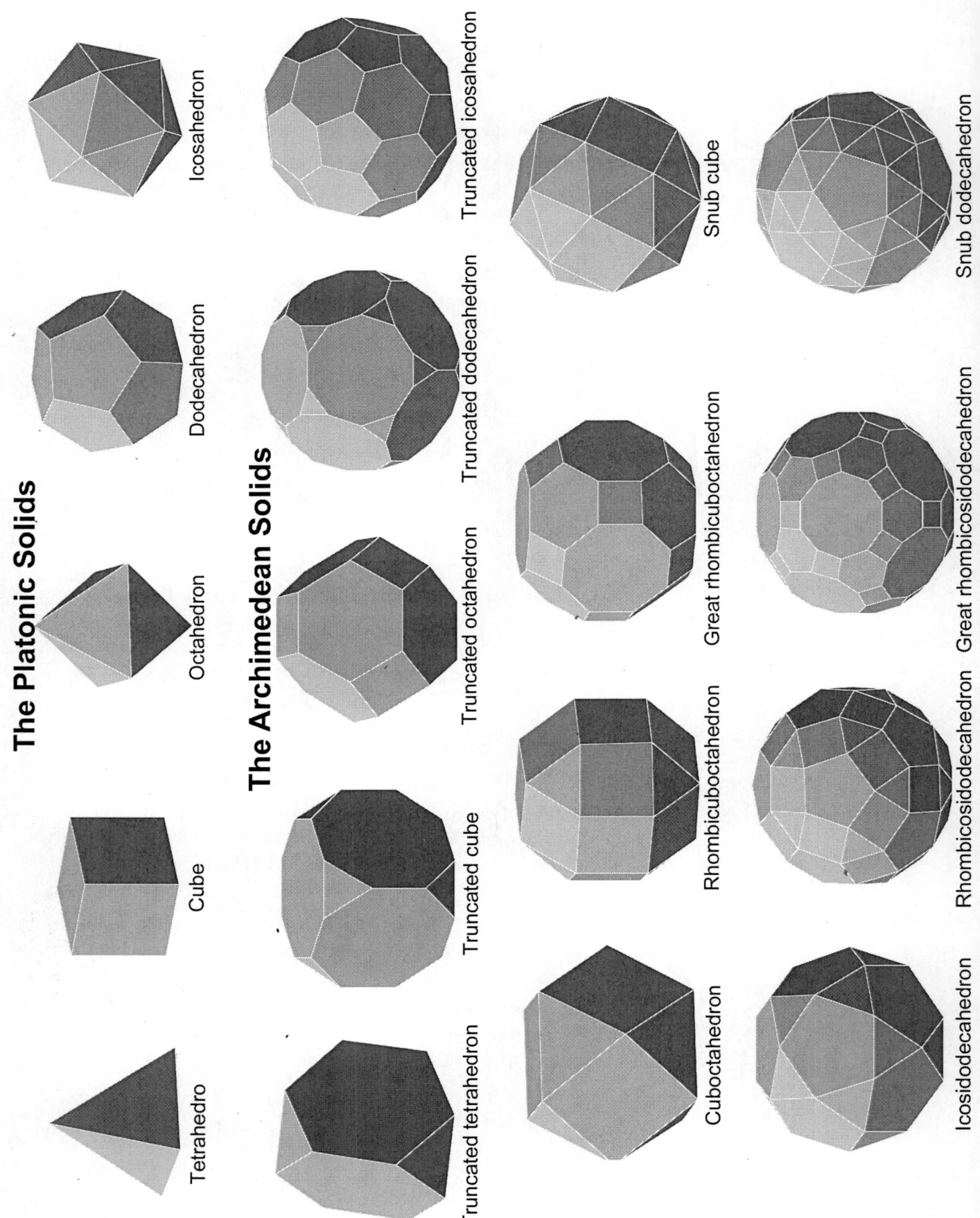

The Platonic Solids

Icosahedron

Dodecahedron

Octahedron

Cube

Tetrahedron

The Archimedean Solids

Truncated icosahedron

Truncated dodecahedron

Truncated octahedron

Truncated cube

Truncated tetrahedron

Snub cube

Great rhombicuboctahedron

Rhombicuboctahedron

Cuboctahedron

Snub dodecahedron

Great rhombicosidodecahedron

Rhombicosidodecahedron

Icosidodecahedron